an Age *of* Barns

ERIC SLOANE'S

AN AGE OF

BARNS

Printed in China
Cover designed by JoDee Turner

02   03   04   05   5   4   3   2

Library of Congress Cataloging-in-Publication Data
Sloane, Eric.
    Eric Sloane's An age of barns / by Eric Sloane.
      p. cm.
    Originally published: An age of barns / Eric Sloane. New York : Funk & Wagnalls, c1967.
    Includes bibliographical references.
    ISBN 0-89658-565-4 (alk. paper)
      1. Barns—United States—History. I. Title: Age of barns. II. Sloane, Eric. Age of barns.
   III. Title.

   NA8230 .S57 2001
   728'.922'0973—dc21

                        2001017728

Distributed in Canada by Raincoast Books, 9050 Shaughnessy Street, Vancouver, B.C. V6P 6E5

Published by Voyageur Press, Inc.
123 North Second Street, P.O. Box 338, Stillwater, MN 55082 U.S.A.
651-430-2210, fax 651-430-2211
books@voyageurpress.com
www.voyageurpress.com

This sketchbook is dedicated to those men who knew America when it was very young and so had the good fortune to know the land first and to realize its worth to the fullest. Their names were seldom recorded, for they were a simple people whose days were spent in doing routine chores of the farm. Each man was woodsman, doctor, carpenter, weaver, farrier, wheelwright, shoemaker, teacher, artist, and—sometimes—soldier. He had a reverence for excellence, a reverence so profound that all things he did, he did well. America's landscape is still dotted with his monuments— the barns and bridges, the farmhouses and mills, built by these perfectionists more than two centuries ago. Few of the structures being erected today will last the lifetime of their makers: many will be replaced as being obsolete, out of date, or not earning enough money to pay their way.

Those who seek the spirit of America might do well to look first in the countryside, for it was there that the spirit was born. While there are still patches of countryside left, and while remnants of the old barns remain, this book will serve as a guide through the past.

We have finally come to realize the beauty and excellence of the homes built by the early Americans, but too often their barns are regarded as mere curiosities. They are, rather, the shrines of a good life and ought to be remembered.

Author's Note

My object in this book is not to tell you how to build or use a barn, nor is it to create an architectural record. My purpose in putting together this album of a vanishing heritage is to remind you of the way people lived in those early days.

In portraying how Americans lived in the past, I do not mean to tell just what they wore or what they did from day to day; instead I want to show how they reasoned and what their attitude toward life was—their personal aspirations and national purpose.

A century or two ago, although a man might be a doctor or printer or lawyer by profession, he was also, of necessity, a farmer. The early farmer who *sold* produce was a rarity; husbandry was not a business but a way of life. A man farmed to feed his own household and his livestock, and closeness to the soil and awareness of Nature were an inherent part of American living. This era has nearly gone now—we are at the end of our "age of barns."

There are still American farmers, of course, and with the population explosion, the farm is more important than ever. But the average small farmer—overworked and underpaid—is not being accorded the dignity that should be part of his heritage. The successful farmer has been transformed into a businessman, and the barn has become a factory.

I find it strange that until now so little has been written about early barns. As I begin this study and sketchbook, interior decorators are buying barn siding and paying more than a dollar a foot for it, and wherever there is a barn worthy of being remodeled into a house, its value has risen high enough to put it among the class of rare antiques. And yet, libraries cannot provide much information about the history or construction of these early barns.

When I first began the research for this book, I went to many libraries in search for old pamphlets or instruction books on building barns. I assumed that when a farmer built a barn, he needed directions. I was surprised to find next to nothing on this topic, and I was greatly amused when some of my own earlier books were recommended to me as guides. (I *knew* they didn't have the information I wanted.) References to barn framing can be found in some old books on architecture, but it seems that the pioneer farmer was mainly on his own when he undertook the actual construction of a barn. This makes it an even greater wonder that old barns from Maine to the Atlantic southland have so much in common.

Our forefathers had a deep respect for tradition and the accepted way of doing things. It was their complete adherence to rules that enabled them to do many things well. Because each person helped with the construction of his own house, I assume that building knowledge was passed from generation to generation. Often, in the crudest early implements there can be found a beading or indented decoration worthy of the most sensitive artisan; in the simplest house framing one can see touches in hand-hewn beams that show a knowledge of classic architecture. Ira Allen, brother of Ethan Allen, wrote of Vermont: "I am really at a loss in the classification of the inhabitants here. They are all farmers, and again every farmer is a mechanic in some way or other, as the inclination leads or necessity requires. The hand that guides the plow most frequently constructs it."

It might be in order at this point to define what the word "barn" meant in those early days. Originally, the word meant "a place for barley." It combined the Old English word *bere* (barley) and *ærn* (place). As a matter of fact, it was a little-used word until after the 1600's. Such lists of "rustick terms" as the *Dictionarium Rusticum* (London, 1681) did not include it. The first dictionary to list it described a barn as "a place for laying up any sort of grain, hay, or straw." But the mention of "a place" included a door-covered hole in the ground, like that mentioned in 1634 by Wood (N. Eng. Prospect 11, xix, 95): "Our hogges having found a way to unhindge their [sc. Indians'] barne doores and robbe their garners."

By 1770 (according to *Kalm's Travels*), the ground storage hole had become a building—the American having originated the idea of putting the farm complex under one roof: "In the northern states of America, the farmers generally use barns for stabling their horses and cattle; so that among them, a barn is both a cornhouse or grange, and a stable."

There are historians who do not agree with my theory that the early American barns are basically *different* structures. A barn is a barn, they reason, with four walls and a roof made in the simplest manner and quite similar to other structures of the time. I maintain that what a man creates is influenced by his reason and purpose for creating it. This is quite evident when you consider why the farmer built his barn the way he did. We so often say, "They don't build houses the way they used to." More to the point, it seems to me, is that now there is *no reason* to build that way!

I once bought an abandoned farm with the intention of remodeling the barn and reviving the farmyard. Although the house had been built in pioneer days, there was a sense of its having been lived in during a more recent yesterday, for at least a dozen families had left their marks. The barn, on the other hand, gave the feeling of never having been changed since the time it was raised.

One night when I was wondering what the original builder might have been doing at the same hour two centuries ago, I chose to think he most likely would have been putting his stock to bed. So, trying to recapture something of the past, I walked toward the old barn. Feeling like a ghost contemplating the business of haunting—there was indeed a powerful sense of another time—I breathed in the winter freshness of the night and felt that I was re-

living some certain moment. I could smell the musty tang of hay and manure, and it was easy to imagine a restlessness of farm animals within the barn. I pushed open the half-collapsed door and stepped into the blackness.

At once, I seemed to have an overwhelming sense of satisfaction and safeness: there was a welcome softness of hay underfoot, and although they could not be seen, the surrounding walls and the oversize beams made themselves felt, almost like something alive there in the darkness. The incense of seasoned wood and the perfume of dry hay mingled to create that distinctive fragrance which only an ancient barn possesses. I felt a strong affinity for the man who had built this barn. Perhaps some of his reverence passed on to me—perhaps that instant was the beginning of my regard and affection for old barns. It takes only an instant for a person to be directed to a path that he will follow for the rest of his life.

Eric Sloane
Weather Hill
Warren, Connecticut

9

an Age of
Barns

PRE-COLONIAL
RANCH HOUSE
\$42,000

Why old Anything?

This is a good question and a timely one. In this jet age, two different schools of thought are rampant, and it is increasingly important that both be recognized. One holds that new things are best, in the main because they are usually some sort of improvement. But those who are irritated with modern times preach that "old ways are usually the best." Of course, there is something to be said for and against each view. New things are not always improvements on the old; often they are sad imitations. And many old things are obsolete—even bad—

and consequently, of no value whatsoever. As a matter of fact, age itself has no value; its only worth is that it provides the time for possible improvement. Age withers, cripples, and finally kills all living things. As a philosopher once wrote: "The only thing about age is that it affords time for learning and for good deeds. If you do not learn or do good works, old age will do for you only what it does for a dead fish—but slower."

Certainly the good things of the past should be sorted out from the bad and rescued from those attic-cleaners who believe only in the new. It is constantly drummed into us by radio and television that anything new is what we should want. As a result, many of us carelessly throw away treasures, both real and spiritual, that took centuries for mankind to acquire.

Buildings that are decaying seem to have little place in our world; even houses still sound but considered "out of date" are eligible for removal. Yet there are buildings in Greece and Rome, certainly in need of repair—some reduced to a few columns and fragments—that still provide standards for architectural beauty. They have become symbols of the way a people once lived and thought; and in their crumbling state, they have acquired an honorable and pleasing decay. The same may be said of some of our early barns.

When costly structures built less than fifty years ago have become obsolete and are being torn down, it is amazing and significant that a simple barn in the country, even in a state of ruin, can continue to benefit and enrich its surroundings after two centuries. And when that barn is threatened by a new housing development, it is usually the old barn that seems attractive, while the new buildings look grotesque—until, of course, the old barn is removed and we can become accustomed to the sameness of mediocre design.

The education of sight—the art of vision—is not being given proper attention today. John Piper in his book, *Buildings and Prospects*, says: "The appreciation of pleasing decay is an important one, because it is so neglected. It is always worthwhile looking at a building twice before pulling it down. A building in a state of pleasing decay should be looked at three times . . . to be sure, first, that it has no virtues in itself that will be sadly missed; second, that it will not be missed as an enrichment of its present surroundings; third, that it might not form

a useful point of focus, whether by agreement or by way of contrast, in future surroundings."

These words are esthetically, morally, historically, and architecturally sound; but to any modern American builder, they are hogwash. For modern values are not based on esthetics, morals, history, or even architecture; rather they are based on profit to be made in a given length of time.

There was once a magnificent old barn that was torn down because a gasoline company wanted the land it stood on and paid a huge sum for it. A local historical society had a plaque made and erected nearby, telling about the barn and its history. But a sense of history is an apology for the absence of beauty; it is in no way a substitute.

The beauty of wood in the state of pleasing decay is one of Nature's special masterpieces. Often a plain board from a barn is so remarkable in its composition—grains and knots and shades of weathered gray—that it could be framed and hung just as any modern painting is. And the old barn, growing moss and lichens, crumbling into decay, can be more a work of art than a new building. It is all in one's manner of thinking. John Ruskin commented on the European acceptance of pleasing decay, saying: "A building of the eighth or tenth century stands ruinous in the open street; the children play around it; . . . no one wonders at it or thinks of it as separate and of another time. We feel the ancient world to be a real thing, and one with the new; antiquity is no dream; it is rather the children playing about the old stones that are the dream. But all is continuous; and the words 'from generation to generation' understandable there."

America has no noble ruins, for the old houses are torn down to make way for the new. But, fortunately, some of the old barns still remain—the only structures that are allowed the dignity of pleasing decay.

The FRAMER'S Implements.

Some barns have been built with nothing but a felling axe. But the great structures were usually put together by men who specialized in framing. The tools they used are illustrated below and to the right.

When the farmer was planning to erect his own barn, he cut the beams at least a year ahead and allowed them to season; he then framed each section or "bent" on the ground and pounded each union into place, using a beetle or commander. Joints were pinned with temporary trunnels (treenails) called drift pins, which could easily be hammered out and replaced with permanent pins after the beams were in place and well seasoned.

Almost all barn beams were shaped with a broad-axe and not with an adze. Dutch barn builders used the adze on some of their more prominent beams, but the well-known "cut-and-sliced" beam shows the cut of the scoring axe and the slice of the broad-axe.

Only wooden hammers and mauls were used by the framer to strike the mortise axes and splitting-wedges. Perhaps his most distinctive tool was the framing hatchet, which he used for many purposes and, therefore, kept hanging always from his leather belt.

When the professional framer was finished, the siding, roofing, and flooring work was always done by less skilled workers.

a Drift Pin →

an Adzed beam

Broad axed

Open handled Saw c. 1750

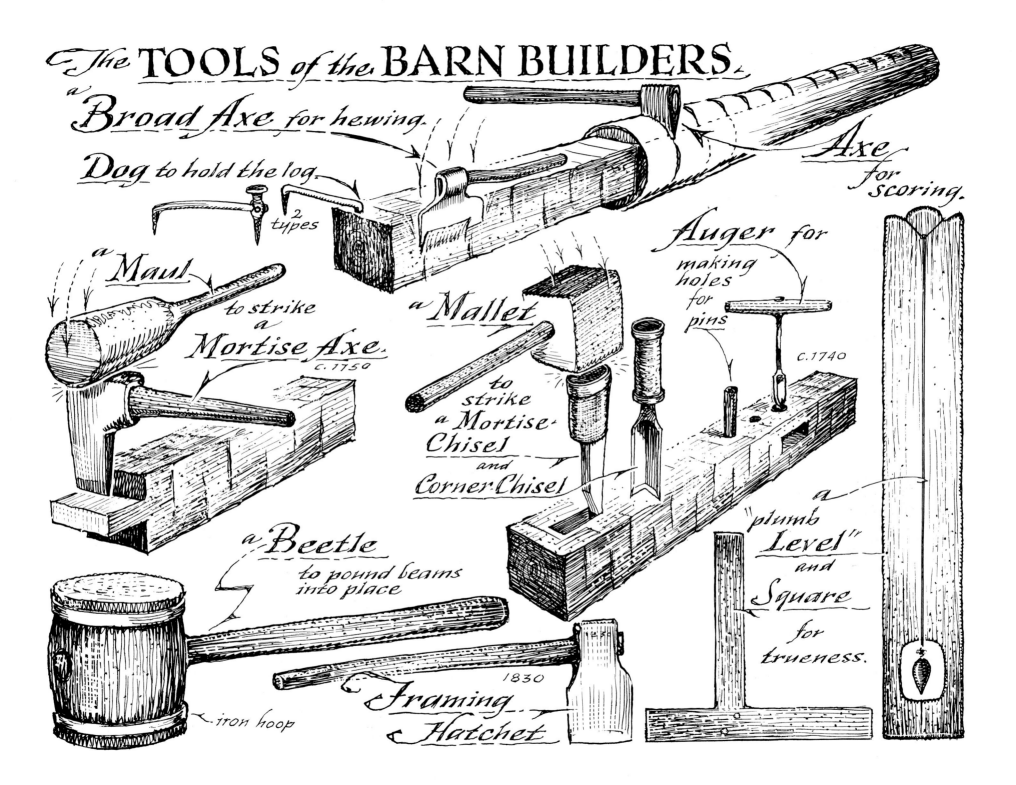

The TOOLS of the BARN BUILDERS

a Broad Axe for hewing.

Dog to hold the log. 2 types

Axe for scoring.

a Maul to strike a Mortise Axe. c.1750

a Mallet to strike a Mortise Chisel and Corner Chisel

Auger for making holes for pins c.1740

a Beetle to pound beams into place

iron hoop

1830 Framing Hatchet

a plumb Level and Square for trueness.

When barns were for storing TAXES.

At right is an undetailed drawing of the European tithe barn. Here you can see some of those barn features which the American pioneer builder chose to discard: the long curved cruck, the curved brace, the strut or "knee," and the long roof that rested its weight entirely on uprights within the barn. Many of the old tithe barns were used for storing contributions to the church; naturally, therefore, their architecture was churchlike, with the entrance at the gable end.

The cruck (also known as crook, crutch, crog, and crock)—a word unknown to our dictionaries—is familiar in England and elsewhere in Europe as a feature of many surviving ancient farm buildings. It was made from one curved tree sawed in half in such a way that each piece matched the other. Then, fastened at the top, it became a giant "A," which was the beginning of the building.

In the Old World, the ancient king post was used only to hold up the peak. In America, however, it was joined into a truss (king post truss) and later enlarged into the double upright truss (queen post truss) that is a feature of most early American barns.

One of the reasons that settlers came to the New World was to escape tithes (or taxes), and so it was natural for them to build their barns in a way different from the familiar tithe barns of Europe.

a Tithe Barn of England c. 1650

Stalls
each compartment called a "Bay."

PEAK COLLAR
KING POST
RAFTER POLE
GIRDER
STRUTS
WALL-POST
STALL POSTS
DIRT FLOOR

MEDIEVAL ENGLISH and GERMAN BARNS

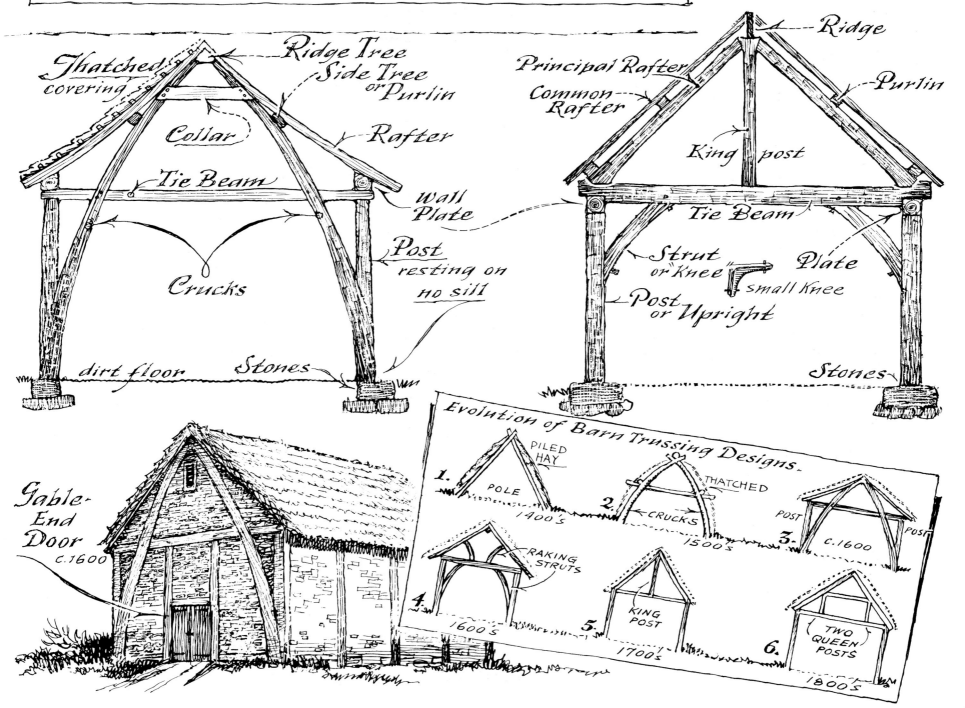

Thatched covering

Ridge Tree
Side Tree or Purlin

Collar

Rafter

Tie Beam

Wall Plate

Crucks

Post resting on no sill

dirt floor   Stones

Principal Rafter
Common Rafter

Ridge

Purlin

King post

Tie Beam

Strut or "Knee"   Plate
small Knee

Post or Upright

Stones

Gable-End Door
c.1600

Evolution of Barn Trussing Designs.

1. PILED HAY   POLE   1400's

2. THATCHED   CRUCKS   1500's

3. POST   POST   c.1600

4. RAKING STRUTS   1600's

5. KING POST   1700's

6. TWO QUEEN POSTS   1800's

1600's 1700's

Birth of the AMERICAN BARN

When the first well-made barns were built in America, they usually had some of the features of their old European counterparts. Their roofs were steep to take thatching, and hay was still kept in outdoor stacks. Siding was horizontal—as it was on dwellings—and it is likely that some builders still used the curved "Ship's Knee" type of braces. But before long, the American barn assumed the compactness indicated in the drawing above. The walls became higher and the roof less slanted. Changing the old custom of having separate places for cattle, sheds for grain, and clay areas for threshing, the barn was made to accommodate everything.

In Europe, the farm dwelling was designed by laying out rooms or areas and then housing them in. The American way was to build the house and then divide it into compartments. The result was a distinctive barn shape.

The interior sketch opposite shows that a full brace was used in the European barn, and there were no sections or "bents" that could be constructed separately and raised later, as was done at American barn-raisings. In the winter months, a hinged board was used to cover the long horizontal window. In American barns, this device was used over the barn door; later, when panes of glass were added, it was called a "door light," and many such barn door lights still exist in New England.

The American Barns of the middle 1600's to late 1600's were European in design...

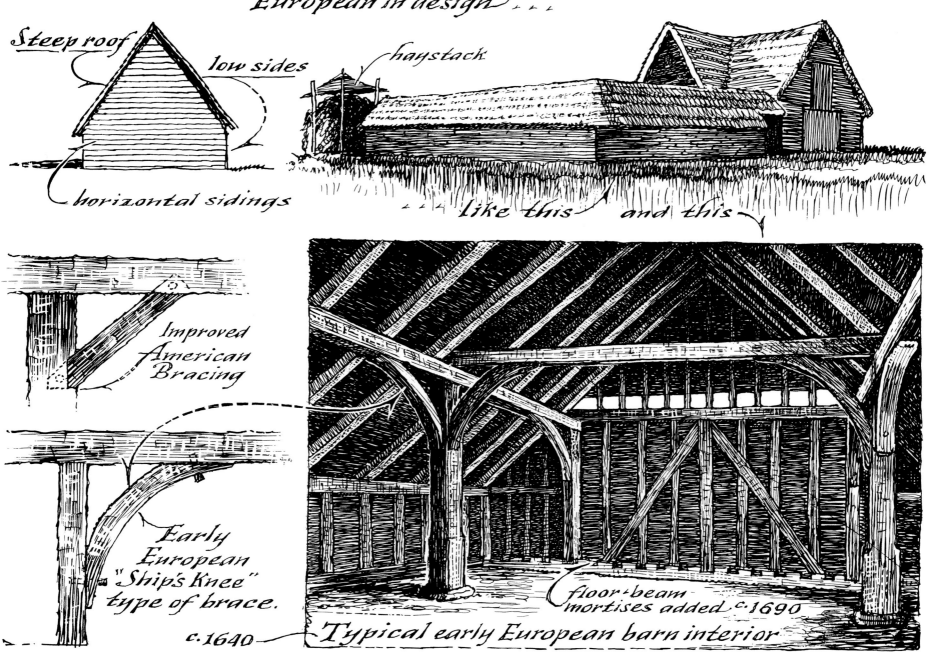

Steep roof

low sides

haystack

horizontal sidings

like this   and this

Improved American Bracing

Early European "Ship's Knee" type of brace.

c.1640

floor-beam mortises added c.1690

Typical early European barn interior

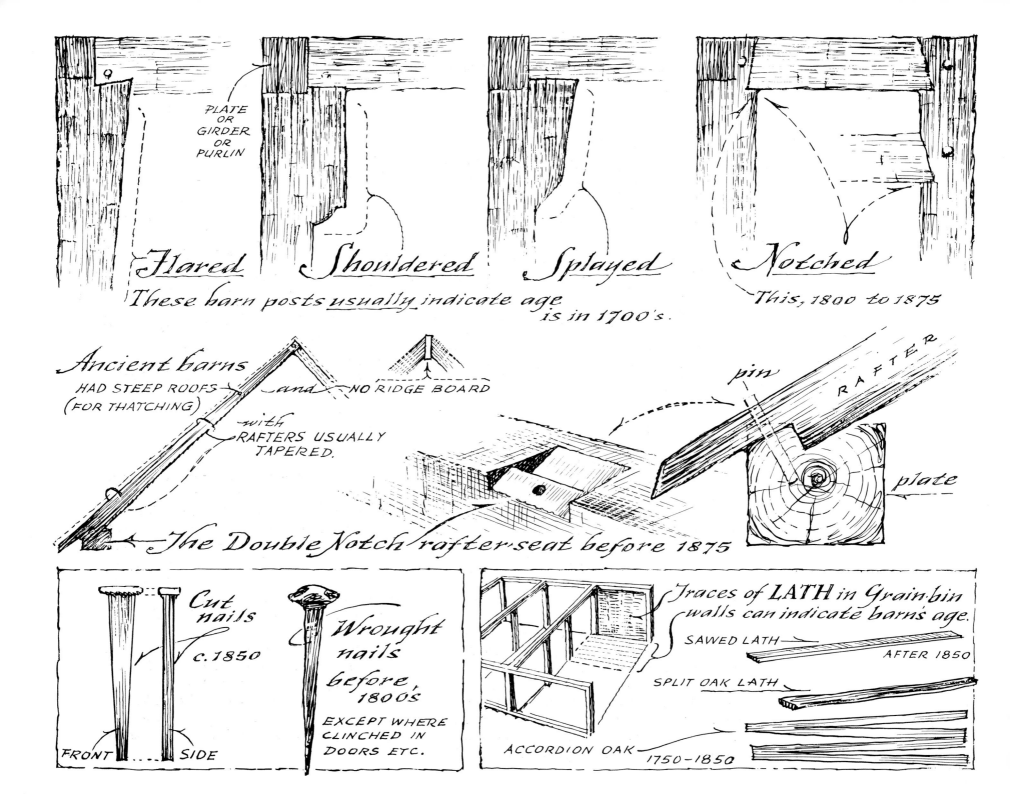

Flared   _Shouldered_   _Splayed_   _Notched_

PLATE OR GIRDER OR PURLIN

These barn posts usually indicate age is in 1700's.

This, 1800 to 1875

Ancient barns HAD STEEP ROOFS (FOR THATCHING) _and_ NO RIDGE BOARD

with RAFTERS USUALLY TAPERED.

pin   RAFTER   plate

The Double Notch rafter seat before 1875

Cut nails c.1850

FRONT   SIDE

Wrought nails before, 1800's EXCEPT WHERE CLINCHED IN DOORS ETC.

Traces of LATH in Grain bin walls can indicate barn's age.

SAWED LATH   AFTER 1850

SPLIT OAK LATH

ACCORDION OAK   1750-1850

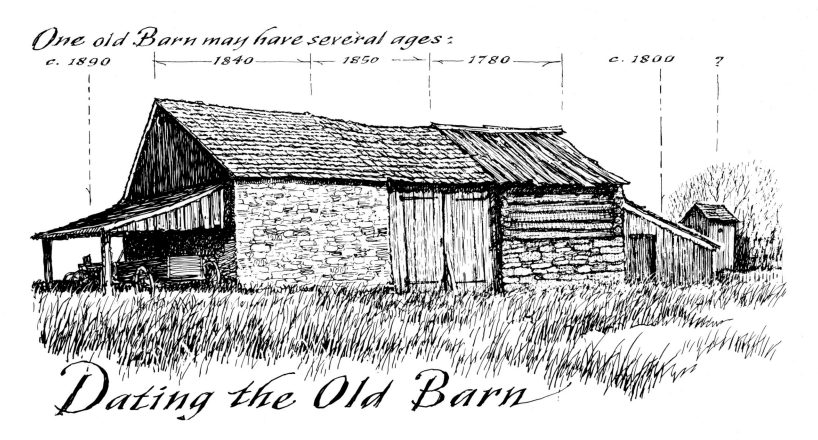

One old Barn may have several ages:

c. 1890   ├── 1840 ──┤├── 1850 ──→│←── 1780 ──→│   c. 1800   ?

Dating the Old Barn

Dating an old, unrecorded barn is a job of detection. Even then, the answer is usually nothing more than an approximate guess and must contain the word *circa*. There are a brave few who feel the need to be definite; and for some reason, they compromise too often and use the date of the year 1798. This, naturally, gives the researcher the impression that most American barns were built in that single year.

Vertical marks of the up-and-down saw usually indicate a pre-Civil War building. And irregularities of those marks may distinguish the older of these structures. Siding put up with cut nails dates a barn as being built after 1800, but wrought nails were used wherever clinching was customary, as in battening doors. Round nails, of course, are recent.

During the 1800's a flat ridge board was devel-

oped; in the 1700's either a heavy five-sided beam was used or there was no ridge piece at all. It may safely be said that most barns of the 1700's had rafters pegged one into the other, with no ridge board.

Remnants of granaries, which were usually plastered to keep out mice and insects, will give you hints of a date in the lathing that held the plaster. Sawed lath came in after 1850; "accordioned" oak sheets were used from 1700 to 1820. The barn above, in three sections, started with the log portion, built about 1780, and the "lean-to," which was added a bit later. The stone addition came into being about 1840, and a few years afterward the two were joined by the part with the door. The shed is recent, but the date of the little outbuilding is anybody's guess. Dating an old barn is an extremely difficult job.

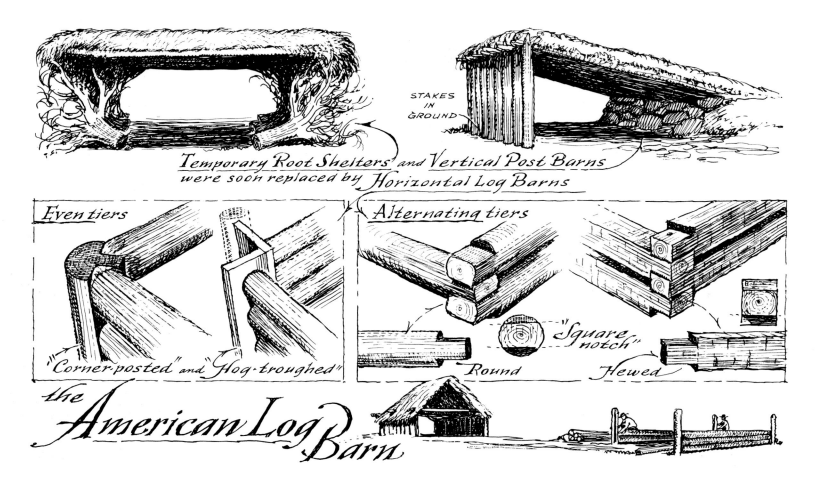

Temporary *Root Shelters* and *Vertical Post Barns* were soon replaced by *Horizontal Log Barns*

STAKES IN GROUND

Even tiers

Alternating tiers

"*Corner-posted*" and "*Hog-troughed*"

"*Square notch*"

Round   *Hewed*

the *American Log Barn*

The log barn is not an American invention, but because the pioneer found a wealth of wood in the New World, log architecture was natural, and soon became complex and varied. Some of the first live-stock shelters were built of tree stumps, which collected and held snow nicely; next came upright logs driven into the ground to create first a stockade and then a wall. Later, upright logs were set between a wooden sill and a plate, and fastened at the corners; this arrangement was favored by French settlers, from Canada to the Mississippi Valley.

The walls of horizontal log barns were laid to tier both *even* and *alternating*. The even-tiered wall was held in place either by mortised corner posts or by a trough of two planks through which the logs were spiked fast. Alternating tiers were favored as being stronger, and soon there were many variations of tenoned cornering.

Only an axe was needed to cut the simple "saddle-notch" and "rough lock-notch" (shown on page 25), but other tightly set cornerings were made with a saw and a wedge. The Germans used a mortise axe. The broadaxe was used to smooth inside and outside walls.

Log barns for cattle were usually left unchinked but sometimes—only during the winter months—the cracks were stuffed with hay. In the early days, farmers favored well-ventilated barns as a shelter from rain and snow. It was thought that a barn that admitted no air caused sickness in livestock.

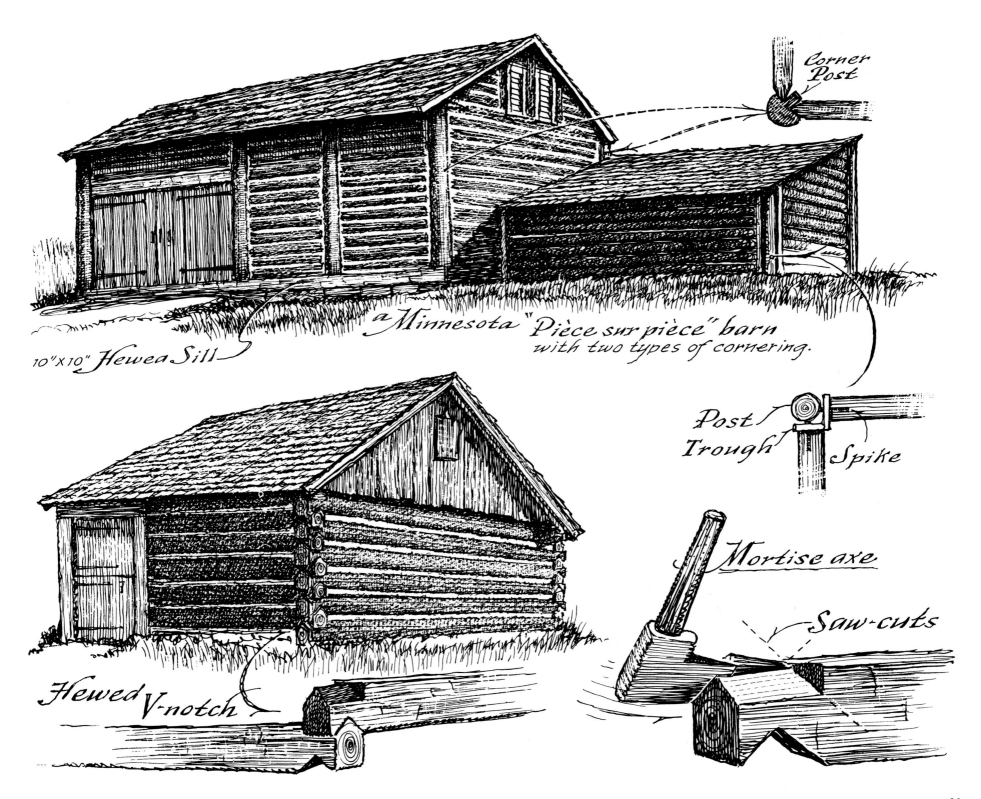

Corner Post

10"x10" Hewed Sill

a Minnesota "Pièce sur pièce" barn
with two types of cornering.

Post
Trough   Spike

Mortise axe

Saw-cuts

Hewed V-notch

The Appalachian Barn began as a Corn Crib and added sheds.

bin

bin

shed

Tennessee
c. 1800

crib / shed

crib / crib

Corn Crib and Gear-shed c. 1875

birth of the Drive-in type,
Double Crib Barn
Virginia c. 1840

a Half-Dovetail Joint

the Small Log Barns of the Southern Mountains

Saddle-notch

Rough Lock-notch

Many of the small Appalachian log barns, with their nicely hewn and chinked inner walls, had first been used as dwellings. Later, after a roomier dwelling was built, the old log cabin became a corncrib, and finally it developed into a full barn.

Corn was the Early American staff of life both for humans and cattle. The corncrib was, therefore, of major importance, and so it developed before the larger barn did. *Crib* meant storage place or enclosure and did not necessarily refer to the slant-walled crib that we know today (a subject discussed later in this book).

The drawings on the opposite page show how the mountain barn evolved from different arrangements of corncribs. The double-crib barn with a connecting shed—the latter had a door—became a full barn when the middle part was used for threshing and the cribs were used as grain mows. It was then that the slant-walled, slatted corncrib came into being.

At first, most barns had roofs of thatched rye straw. Later they were shingled or "shaked."

By the end of the eighteenth century, there were about nine thousand taxable log barns in Pennsylvania alone. Less than two thousand barns were made of stone; and it is presumed that about half of all American barns in existence during the Revolutionary period were built of logs.

A Four Crib Barn

The Big Log Barns

The log barn above was built on four log cribs with a crossed wagonway running through the center. The upper part—almost a separate building—was used for hay storage. This was an English design, and there were many such structures in Tennessee and the Carolinas.

The German version of this barn arrangement incorporated cantilever logs to create an overhung loft (as shown at the top of the page opposite). They were usually built on two cribs instead of four. Eventually, the German design was modified so that there was only one overhung loft. This was the first Sweitzer-type barn, and in it the cattle were below and mows were above. The overhang was on the southern side, and there was either a hill or a ramp on the northern side. This type (also shown on the opposite page) was called a "log and stone

barn" because the log barn, according to old tax terminology, was "without foundation or simply resting on corner stones."

The first story of the Sweitzer bank barn was built entirely of stone. In later barns, the stonework was continued up the two gable sides, while the north and south sides were of log and wood. Finally, in the construction of the Pennsylvania bank barn, logs were completely discarded, and the well-known Pennsylvania stone barn was born.

The weight of the log barns kept them in place without any need for internal bracing, and it is a tribute to the builders that these barns remained true for more than a century. There still exist some large two-story bank barns with wood siding and stone that actually cover the original structure of logs.

Appalachian Overhung-loft Barn
built on two cribs.

crib   crib

corn
crib   shed   mow   mow

Pennsylvania Log Bank Barn
c. 1800

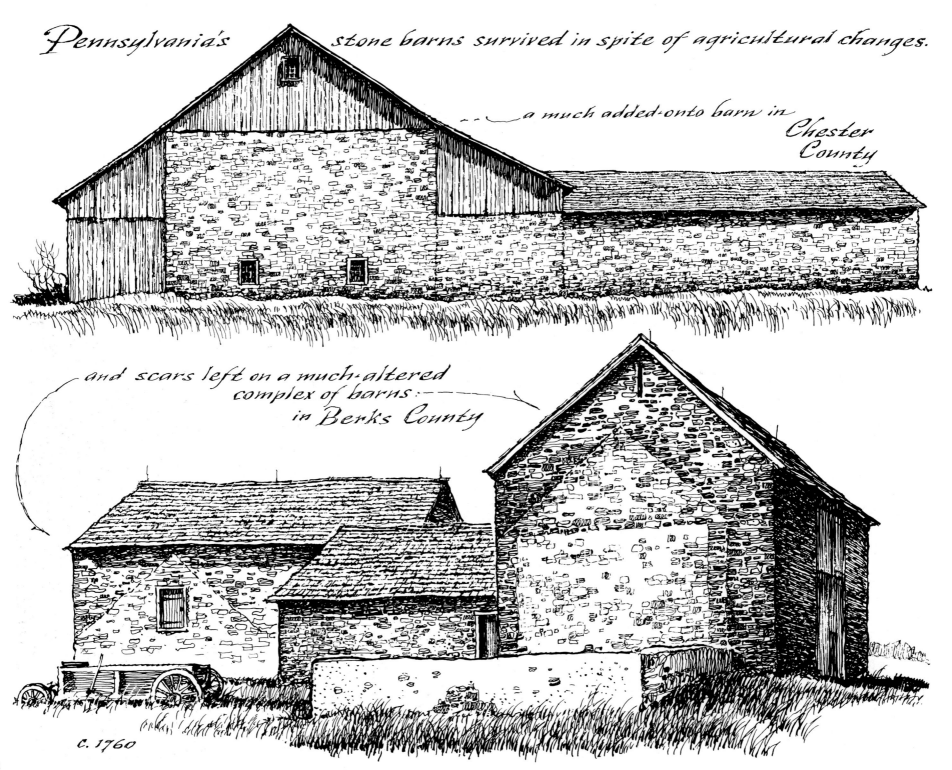

Pennsylvania's stone barns survived in spite of agricultural changes.

... a much added-onto barn in Chester County

and scars left on a much-altered complex of barns in Berks County

c. 1760

Pennsylvania had no _hex signs_. It was all done
"Chust for pretty".

...starting
about
1830

29

the Bank Barn

The Sweitzer barn (also called Swisser or Switzer barn), with its cantilever "forebay" and its banked animal stalls below, reached the peak of its popularity in the late eighteenth century—after about fifty years of development. The typical bank barn began as a sheltered mountain structure with a protecting overhang "on the lee side of the winter" and stall doors facing the south, like the ones shown in the illustration at the right.

The barn shown on the next page is a "single-decker" and is relatively small when compared to some of the palatial "double-deckers" of the early 1800's. Notice that the entrance ramp is flanked by two dry cellars that are partially sunk into the hill, each with a tiny window. One was used for horse feed and the other for potato storage. There was a third cellar, even better protected, that was used for the storage of turnips.

Hay was thrown down the stairway or into the straw room, and here it was fed to the cows. Overhead, on the barn floor, were two threshing areas and two granaries; these were situated in each corner of the overshoot.

Many variations of the bank barn evolved, all typical of Pennsylvania architecture and reminiscent of the time when the farmer was king and barns were the palaces of America.

The Sweitzer barn in America, started small, and as a log structure.

Cornstalks and Clapboard insulation c. 1720

A full forebay

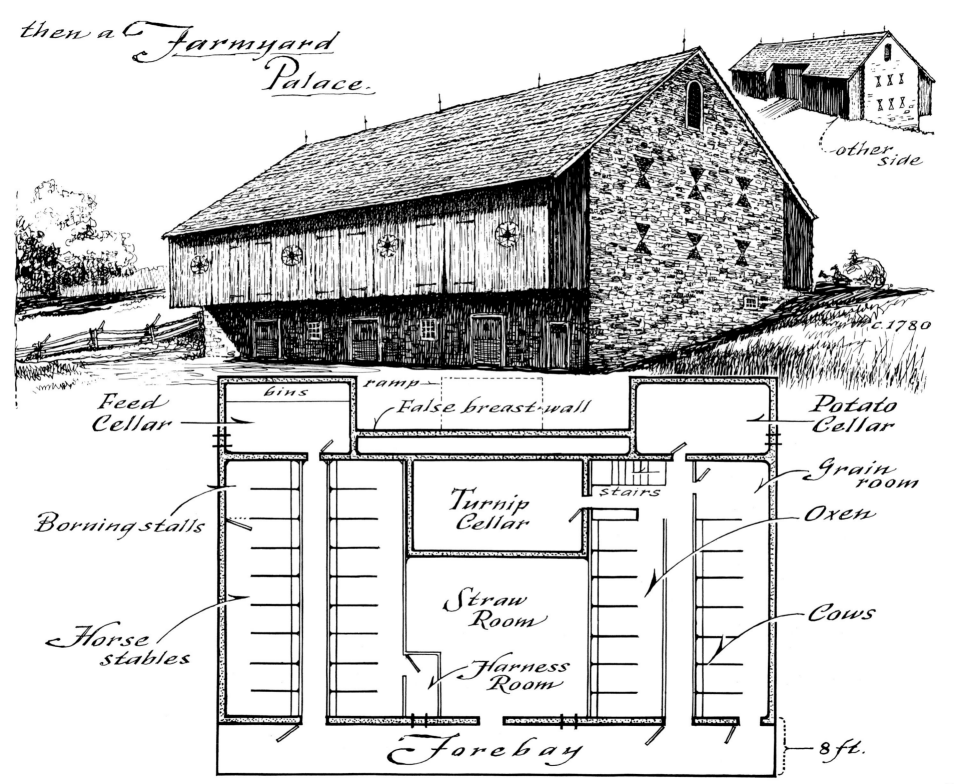

then a Farmyard
Palace.

other
side

c. 1780

Feed
Cellar

bins

ramp

False breast wall

Potato
Cellar

Grain
room

Borning stalls

Turnip
Cellar

stairs

Oxen

Horse
stables

Straw
Room

Cows

Harness
Room

Forebay

8 ft.

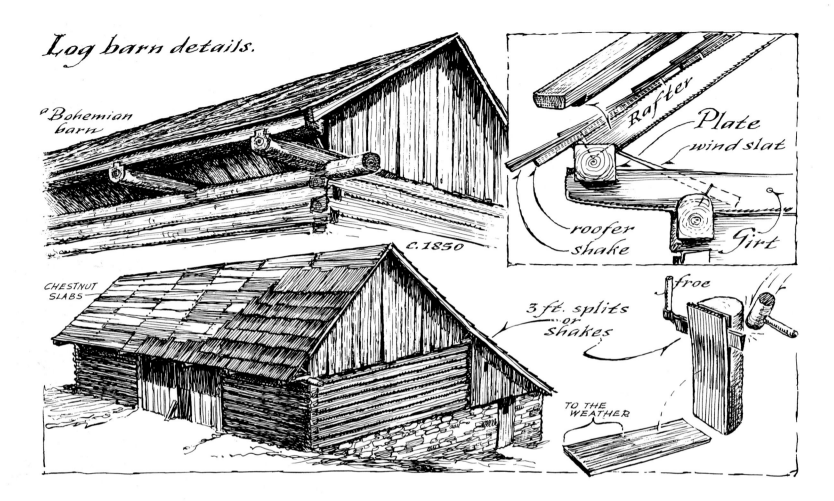

Log barn details.

Bohemian barn

c. 1850

CHESTNUT SLABS

Rafter

Plate

wind slat

roofer shake

Girt

froe

3 ft. splits or shakes

TO THE WEATHER

Early shingling methods were far from efficient. Slabs of wood from three to four feet long were overlapped so that one-half of the shingle was covered and one-half was exposed to the weather. Nowadays shingles are half that size, and we cover two-thirds, leaving only one-third of the shingle exposed. On small roofs, overlapping slats were placed from the eaves to the peak and were called "full-length shingles." Scooped logs and boards purposely warped for this use were sometimes placed over and under—as tiles are—but they all leaked to some extent, so sod or moss was often added to keep the rain out and the roofing in place.

Riven slats or narrow clapboards were nailed over the open spaces between logs, and slats fashioned like shingles kept the wind from entering under the eaves. The generous overhang of the Bohemian log barn kept out bitter weather.

The Georgia barn has clapboard slats along the peak and also covering the log wall interstices. Such log barns were sturdy, but what with mud chinking and split weather-stripping, the gentleman farmer, at the beginning of the 1800's, wanted a more refined and better-looking style of barn architecture.

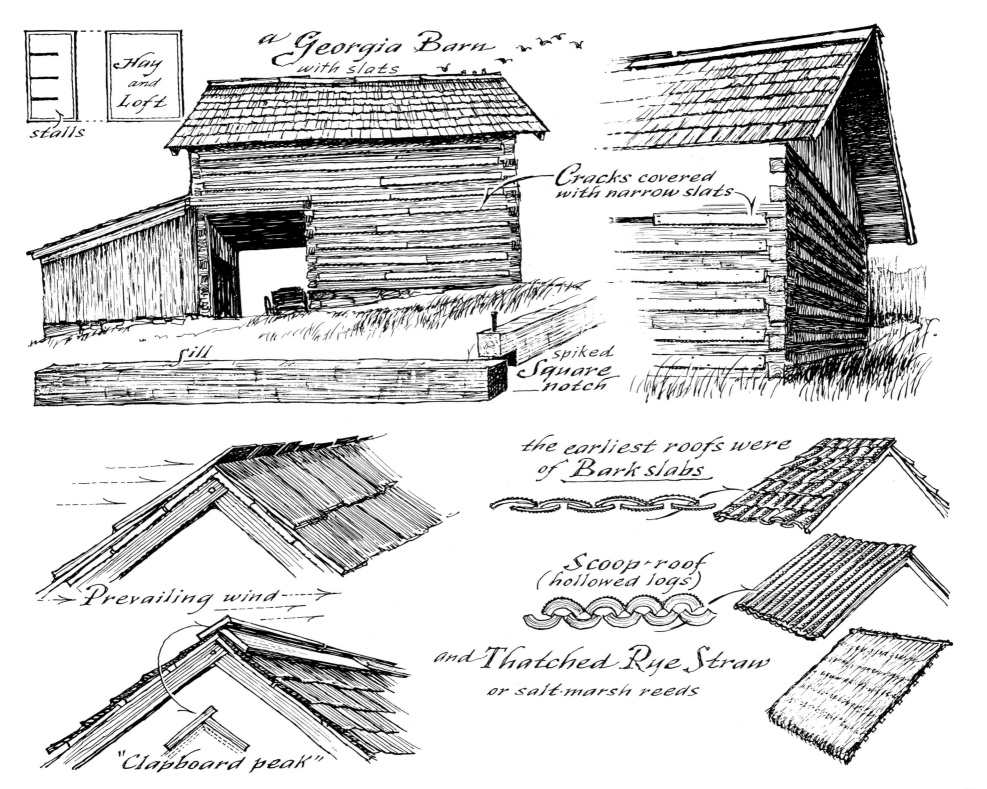

stalls

Hay
and
Loft

a *Georgia Barn*
with slats

Cracks covered
with narrow slats

sill

spiked
Square
notch

the earliest roofs were
of *Bark slabs*

Prevailing wind

Scoop-roof
(hollowed logs)

"Clapboard peak"

and *Thatched Rye Straw*
or salt-marsh reeds

Sawed Siding

cut by a *Gang-saw*

(On early siding,
you can still see the up-and-down
marks on the unweathered
side.)

5

the earliest

Riven
Clapboards

were split with
a *Froe*, which
was struck
with a *Maul*

AND
TWISTED
"TO
AND
FROE"

THE

CARPENTER's

Pocket Directory:

CONTAINING

THE BEST METHODS OF

FRAMING TIMBER BUILDINGS

OF ALL FIGURES AND DIMENSIONS, WITH THEIR SEVERAL PARTS,

As Floors, Roofs in Ledgment, their lengths and Backings; Truſſed Roofs, Spires, and Domes; Truſſing-Girders, Partitions, and Bridges, with Abutments; Centering for Arches, Vaults, &c. cutting Stone Ceilings, Groins, &c. with their Moulds; Centers for drawing Gothic Arches, Ellipſes, &c. &c.

WITH THE PLAN AND SECTIONS OF A BARN.

Engraved on Twenty-four PLATES, with EXPLANATIONS, forming the moſt complete and uſeful work of the kind yet publiſhed.

BY WILLIAM PAIN.

Architect and Carpenter, Author of the Practical Builder, the Carpenter's and Joiner's Repoſitory, and the Britiſh Palladio.

PHILADELPHIA:

PUBLISHED BY J. H. DOBELBOWER AND J. THACKARA.

1797.

PREFACE.

IN the art of building, an intelligent and expert carpenter is entitled to the foremoſt place, or firſt degree of eminence; for he is able to erect a houſe without calling in either bricklayer or maſon, neither of whom can execute the like taſk independent of his aſſiſtance. His profeſſion depends on the practical application of the moſt plain, ſimple, and unerring principles; and more pleaſure reſults from the view, as well as more comfort from the uſe, of a neat well-conſtructed common houſe, than from the moſt ſupurb but ill-contrived palace, where fanciful ornaments are frequently introduced with no better intention than to diſguiſe blemiſhes in proportion and ſymmetry. Strength and convenience are the two moſt eſſential requiſites in building; the due proportion and correſpondence of parts conſtituting a beauty that always firſt attracts the eye; and where that beauty is wanting, carving and gilding only excite diſguſt. In like manner, the affectation of gaudy dreſs in a man who has the misfortune to be deformed, anſwers no other purpoſe than to invite ridicule.

Architecture being founded on geometry, a carpenter ought to qualify himſelf in the fundamental principles of that ſcience: he ſhould acquire a proper judgment of the materials he works on, both as to quality and quantity, and in the uſe of them; the dimenſions of ſcantlings according to the bearings; the truſling of girders and partitions, and the dove-tailing of collars. To aſſiſt the young practioner in theſe particulars applicable to common buildings on given plans and elevations, under the ſeveral articles of framing, flooring, roofing, ceiling, and arching; with the conſtruction of bridges, groins, and ſpires; is the purpoſe of the following ſheets. Every object treated of, is ſo circumſtantially and accurately repreſented in the plates, that very little explanation is needful to comprehend them; the deſcriptions are therefore proportionably brief, to avoid confuſing the mind by a multiplicity of words.—In concluſion, the intention being to concentre every material operation in the common branches of carpentry; it is hoped nothing is omitted neceſſary to enable an attentive workman to become his own ſurveyor, and to execute whatever inſtructions he receives, on true principles, according to the moſt correct methods.

a Book of Plans

Log barns remained popular longer than log dwellings. Both were inexpensive and efficient, but logs were looked upon by many as a crude material, good enough only for pioneers. People, therefore, began to cover their log barns with siding and riven clapboards, and they found the result good to look at. Consequently, the log barn began to lose its popularity. The gang-saw was satisfactorily cutting pine suitable for siding, and instructions for framing barns began to appear in building manuals—such as the one shown above. The colorful language and the old typefaces make the reader instantly aware of an early day, yet the principles discussed are worthy of study by any modern builder. These, and the ones reproduced on the following pages, are from the original book, which is in the author's library. Because of damage due to age, they had to be retouched slightly, but other than that, they are true reproductions.

The *Preface* proclaims the keynote of early barn building: "Strength and convenience are the two most essential requisites," while unnecessary devices only "excite disgust" and "invite ridicule."

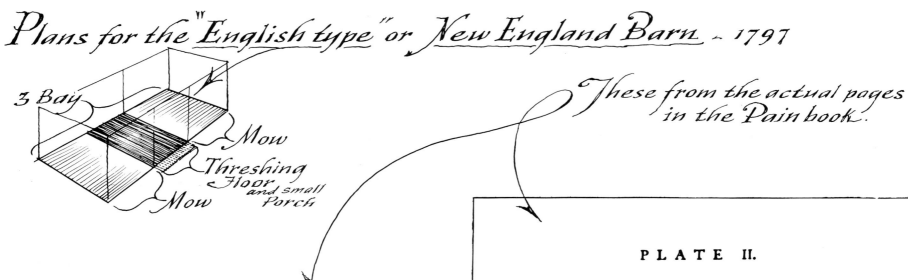

Plans for the "English type" or New England Barn – 1797

3 Bay

Mow

Threshing Floor and small Porch

Mow

These from the actual pages in the Pain book.

PLATE I.

The Elevation and Part of the Plan of a Timber-built Barn.

SHEWS the manner of framing the timbers together, with a porch at entrance. The meafures of the timbers are all figured for the fcantlings in common practice. The camber of the beams to be 1 ½ inch. The whole length of the building 42 feet from out to out; and the breadth 28 feet. Thefe fcantlings will do for a building twice or three times the length, but of the fame width, or not exceeding 30 feet wide.—If the building fhould exceed the foregoing height and width, the fcantlings muft be in proportion to their length and bearings, as in the table of timbers.

PLATE II.

Fig. A is the Elevation of the End, and Fig. B the Section of the Barn, with all the Meafures of the Timbers figured for Practice.

THE whole width from out to out is 28 feet, and 26 feet 4 inches within, with gable-ends for more convenient room. The dotted lines in the fection of the roof reprefent ftretching pieces to go between the purlines, to prevent them from fwagging down, the bearings between the principals being near 13 feet long. As the meafures are all figured, they will appear plainly on infpection. The ground fills to be 10 inches by 6, the main poft 8 inches by 10, the door poft 8 inches fquare. The interftice in the fides, between the poft, to frame the quarters or punchings in, 8 inches by 8, braces 6 inches by 4, punchings 5 inches by 3, raifing plates on the top of the poft 8 inches by 6; braces there from the poft to the beam 9 inches by 3, teffel or top part of the poft 1 foot 3 inches by 8 inches, beam 11 inches by 9, king-poft 1 foot 1 inch by 6 inches, the fhaft of ditto 6 inches by 6, when the butments are cut ftraight to the king-poft; the rafters 4 by 4; the principal rafter at bottom muft be 9 inches by 5, at top 7 inches by 5.

The purlines or girt pieces framed between the principal rafters to be 6 inches fquare, for the reception of the fmall rafters, and to be framed into the principal rafter horizontally, or level with the bafe-line, as reprefented in the fection; the fmall rafters 3 inches by 4 ½.

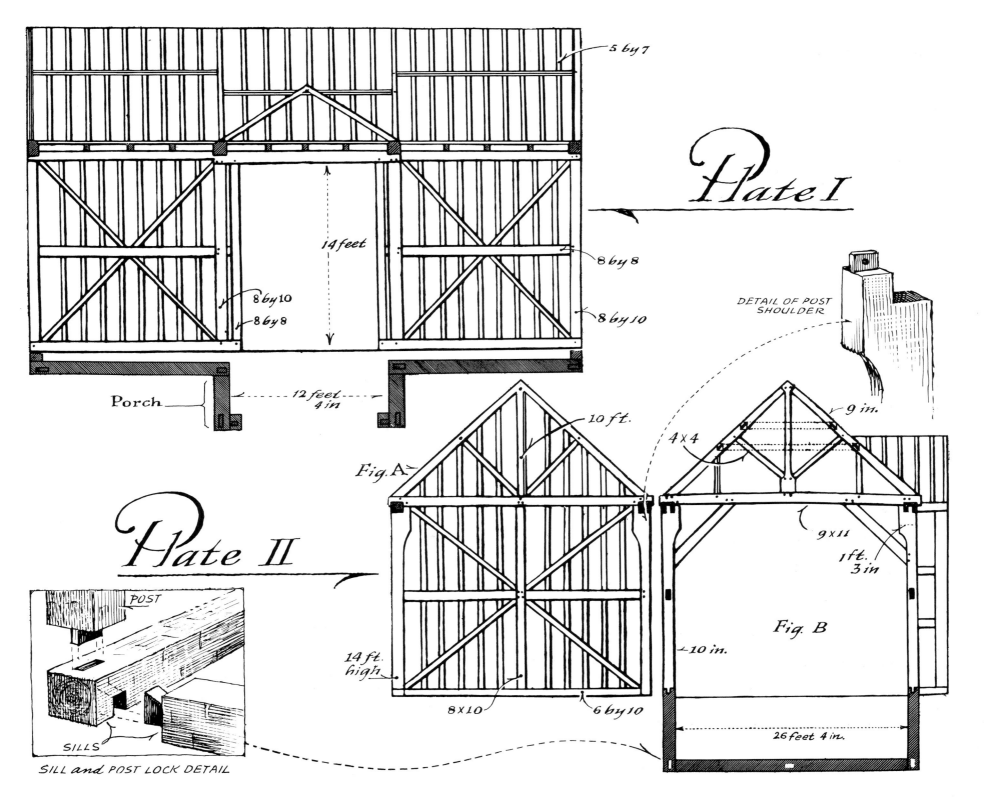

Plate I

5 by 7

14 feet

8 by 8

8 by 10

8 by 8

8 by 10

Porch

12 feet
4 in

DETAIL OF POST
SHOULDER

9 in.

10 ft.

Fig. A

4 x 4

9 x 11

1 ft.
3 in

Plate II

POST

Fig. B

10 in.

14 ft.
high

SILLS

8 X 10

6 by 10

26 feet 4 in.

SILL and POST LOCK DETAIL

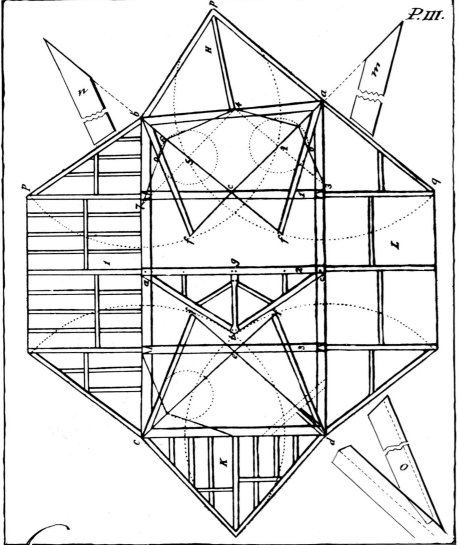

PLATE III.

The Plan of a Roof in Ledgment, shewing the Method to find the Length of the Hips, Square or Bevel, and their Backing to any Pitch required.

LET *a, b, c, d,* be the angles or corners of the building, to find the length of the hips and their backing.

First, lay down the plan of the roof *a b c d* to a scale of one inch to a foot, as the scale *a b;* then, according to the ground you have to build on, take your dimensions, and draw the plan: supposing the plan *a b c d* that to be roofed in; then draw the principal rafters on the plan *a b c,* and dispose of the beams at proper distances, as room will admit, which beams, numbers 1 and 3, will stand to receive the top of the hips; then draw the base lines of the hips *a—c, b—c,* at the bevel-end; and at the square end draw the base lines *c—e, d—e;* then take the perpendicular height of the principal rafters *g, b,* and set it perpendicular from the base lines of the hips *a c, b c,* and *c e, d e,* as *c f* and *e f;* then draw the lines *a f, b f, c f, d f;* these lines will be the length of each hip respectively. Then, to find the backing of the hips, draw a line square with the base-lines of the hips, as 4, 1, 3, and 4, 5, 7; then set the compasses at 1, and extend them to touch the hip at *o,* and draw the small dotted circle as there described; then from the point 2, where that circle cuts the base-line, draw these lines 2—3, 2—4, which are the backings of the hips: proceed in the same manner at the angles *b, c,* and *d,* as will appear plain to every practitioner on inspection.

H, I, K, and *L,* are ends and sides in ledgments; *m, n, o,* are the bevels for the feet and tops of the hips to lay out the sides and ends. Set the compasses at the angle *a,* draw the dotted circle *p f q;* then set the compasses at *b,* draw the dotted circle *f r p,* which gives the length of the hips and rafters laid out. This is a general method for finding the length and backings of hips in any case required, square or bevel, which will plainly appear in all the following plates. Notwithstanding the plans may be different and irregular, the method is the same in every respect.

The Hip Roof Barn

Virginia, 1760

...an early American show-piece with its complications.

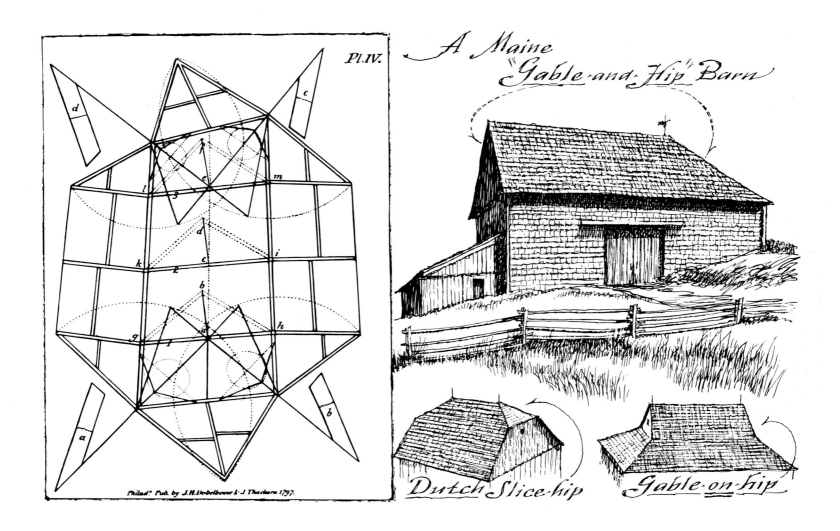

PL. IV.

A Maine "Gable-and-Hip" Barn

Philad.ᵃ Pub. by J.H. Hobelhoure & J. Thackara 1797.

Dutch Slice-hip

Gable-on-hip

Most of the more stately early dwellings had hip roofs, and when a gentleman farmer wanted to build an exceptional barn, he chose this design. But the problems of matching rafters and raising the frames in a hip-roof building were not for the average farmer. The slightest irregularity in the shape of the barn affected the accuracy of hip-roof rafters unless the builder used a compass and followed directions carefully to determine the irregular lengths and cuts of each rafter.

Plate IV (above) looks like a builders nightmare,

because all four walls of this barn are irregular. Mr. Pain included this plate in his manual to show the procedure used in determining rafter proportions despite any irregularities in the shape of the building.

In New England, there are still some stone barns with hip roofs, but few of the frame hip-roofed barns remain in America.

The gable-on-hip was popular as a tobacco-barn roof, and even today many of those can still be seen throughout the South.

Pentroofs

Pennsylvania

a "Watertable" is a projection to throw off water

a Shake Pentroof

Water-table

Pent-roof

Today's common definition of the word "penthouse" is far from the original, which was taken from Old French *(apentis)*. Then a penthouse was not a house or an apartment but only a small roof attached to a building. Even my use of the word "pentroof" on the sketch above is not entirely correct, but it is now accepted barn terminology.

The sketch is of an ancient bank barn. According to old Pennsylvania custom, floor joists were allowed to protrude beyond the masonry to support a narrow pentroof—usually built along the southern wall. The pentroof of the larger, older building has rotted and disappeared, leaving the telltale mark of a water table, and showing still the stubs of the joists. The pentroof on the smaller and newer wing is still intact, although warped and decaying. If you see wood protruding from the inner structures of a barn, you can be pretty sure it is evidence of some long-gone appendage. And the antiquarian detective can safely assume that there was once a pentroof on the building.

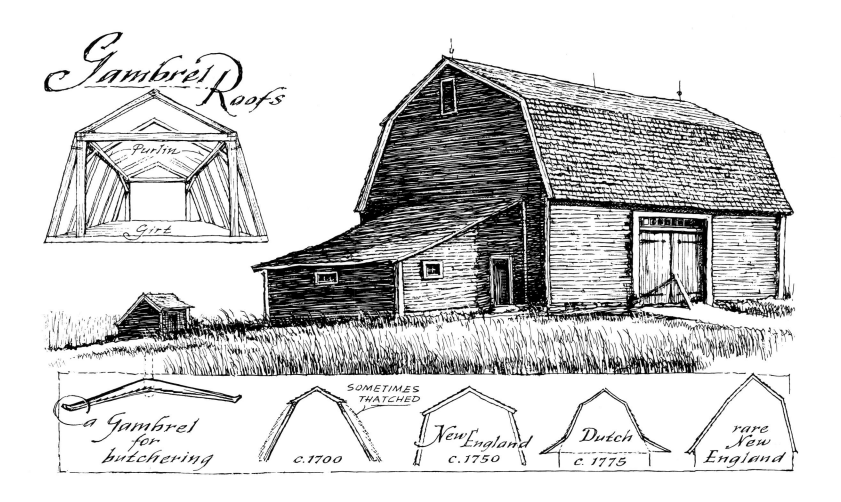

Gambrel Roofs

Purlin

Girt

a Gambrel for butchering

SOMETIMES THATCHED

c. 1700

New England c. 1750

Dutch c. 1775

rare New England

In most old barns, you will find one or two "gambrels." These were used for the hanging of carcasses at butchering time. The word derives from the hock (or bent part) of a horse's leg, which is also called a gambrel—as is the bent rafter line of the roof.

The purpose of the gambrel roof is to make the most use of the space within the roof area. Many examples are conversions of plain peaked roofs, and often only one side has a gambrel while the other side has a lean-to. The Dutch gambrel (shown in exaggerated form) had a curve at the eaves where the rafter was joined with a "Dutch knuckle" extension; this was used for small barns and farm outbuildings.

A gambrel roof on a barn usually indicates a European influence; the one shown above was built by a German. Notice the door light, which was once an opening equipped with a hinged board, but is now glazed. The barn door opened onto a threshing floor, with mows on either side.

Extended Bays

The division between any two framings in a barn is called a "barn bay." An open space between the floor and the wall—for pushing hay to the level below—is called a "hay bay." And extensions that enlarge the interior—in the manner of a bay window—are called "extended bays." A few examples of extended bays are shown in the illustrations. The Dutch grain loft bay and German barn door bay (right) protrude only a foot or two from the main building. The threshing floor of the Pennsylvania barn (opposite page) extends out and over the protruding water table, which can be seen on either side of the bay. The water table indicates that a pentroof was once attached to the barn.

In the Ohio example, you will see that extended bays are often used for storing grain in an effort to keep it dry. The Connecticut barn's "covered bridge bay" affords storage space below and gives sheltered access to a root cellar.

When the extended bay becomes a complete addition to the barn, with a continued roof (for example, the Pennsylvania overshoot), it is called a "fore-bay." As these are usually erected on cantilevered joists, they are not added extensions, but were built as part of the original barn.

Barn Bays

a Hay bay

an Extended bay

Dutch barn
New York
c.1780

grain loft

Lancaster County
Pennsylvania

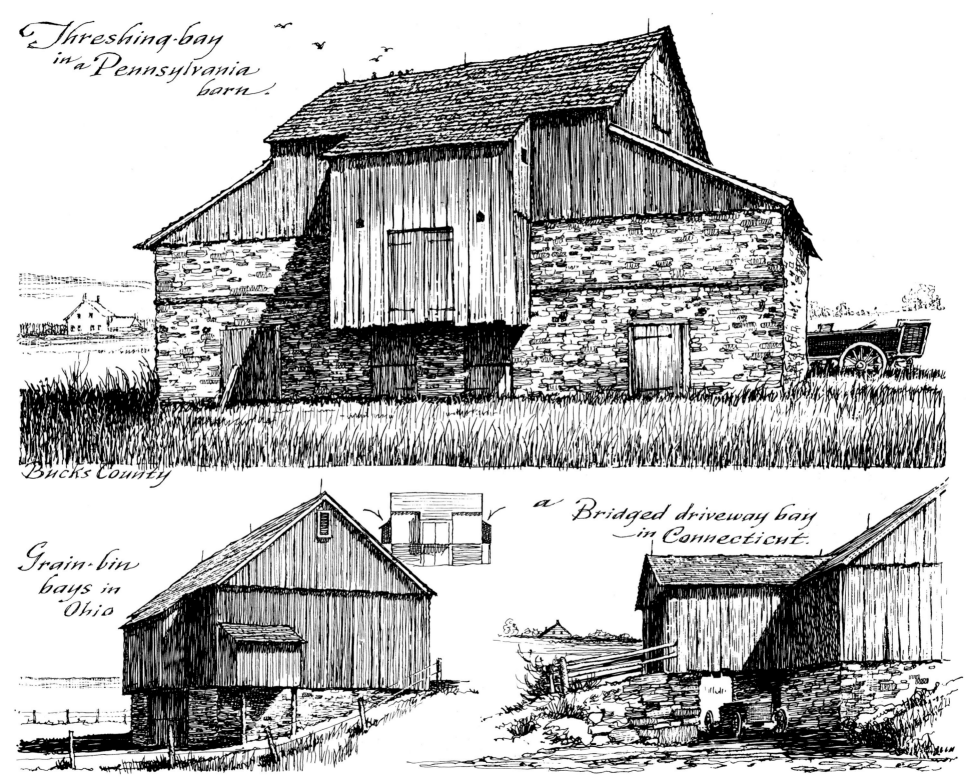

Threshing-bay in a Pennsylvania barn.

Bucks County

Grain-bin bays in Ohio

a Bridged driveway bay in Connecticut.

43

a Salt-box
Barn

Tennessee
c. 1860

3   1   2

New Jersey
c. 1840

Additions made salt-boxes too.

Architectural historians frequently insist that the American saltbox design was really an afterthought or the result of an addition to a barn. True, this was very often the case. But there were also some early houses purposely built in pure saltbox design so that a long protective roof could face the north and the higher, exposed portion would face the warm south. This was an American idea, evolved to cope with American weather, and it seems to have been developed almost entirely by the pioneers who came from England. The very small saltbox barn appears often in the South; it is quite evident that the long sloped roof was not an afterthought, because the strongest wall was always the inner one (marked with an X in the drawing on the opposite page), and the rafters were in one piece. The Germans often made additions or lean-to wings on their barns, but the addition was usually built toward the south—a sort of "reverse saltbox" (see drawing above).

The few barns of true saltbox design that remain today, like the little log barn shown above, hark back to the time when our pioneers were beginning to develop American designs.

Continuous Architecture

This rambling design very likely got its start in New Hampshire, when some farmer broke through the wall of his attached woodshed to make an opening into his kitchen. At any rate, the high snow was responsible for New England's continuous architecture. There are hints of connected buildings in other parts of the country—wherever a washhouse or summer kitchen is connected by a breezeway to the main house—but only in New England do you see the complex of farm buildings that can truly be called "continuous architecture."

Barns never spread out from both sides of the farmhouse; instead the buildings wandered in one direction only (or sometimes in an L shape). As a result, a whole day's chores could be done sheltered from bad weather.

In the 1600's, continuous barns were banned by some New England villages as being fire hazards, and a fine was to be levied against anyone who went against the ban, but there is no record of anyone ever having paid such a fine. In the 1700's, the ban was dropped, and it became the farmer's own business if he wished to connect his buildings and thus create a fire hazard.

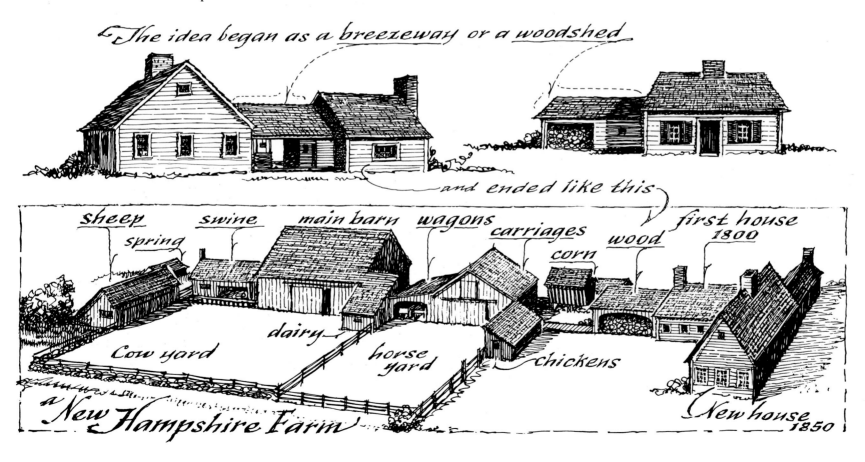

The idea began as a breezeway or a woodshed

and ended like this

sheep · spring · swine · main barn · wagons · carriages · corn · wood · first house 1800 · dairy · Cow yard · horse yard · chickens

A New Hampshire Farm

New house 1850

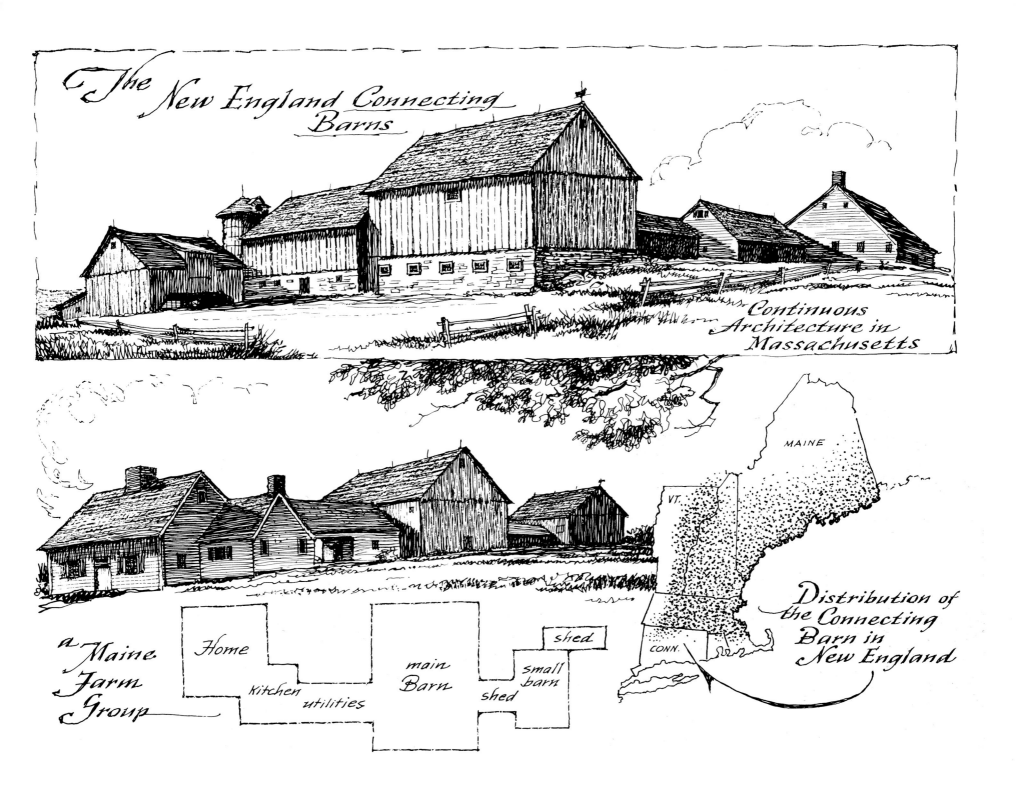

The New England Connecting Barns

Continuous Architecture in Massachusetts

a Maine Farm Group

Home

kitchen   utilities

main Barn

shed   small barn

shed

MAINE

VT.

CONN.

Distribution of the Connecting Barn in New England

47

Heart of the Barn

The typical early American barn was built around a threshing floor, with mows on both sides and doors at both ends of this threshing floor. A waist-high mowstead or threshing wall divided the mows from the threshing area, and each mow had a ladder that was used for climbing into the storage pile. When livestock was kept below, the stairway was near the main door, and hay was tossed down this stairwell or through openings in the floor—the "hay bays."

The grain bin or granary was placed at the end of one mow, often in an overshoot or cantilevered bay to keep it high and dry. It was usually plastered and equipped with a fine door, and the hay mow extended over the granary roof. Opposite the granary, in the overshoot, was a place where threshed grain (both kernels and chaff) was stored prior to winnowing or separating; this area was called a "cove" or "cupboard."

There had been a time when grain was horse-trodden, combed by hand, or crushed by sledges, but the accepted American way was by flailing. Oats, beans, corn, wheat—each had a special flail. The flail was grasped in both hands, the club end or swingle came down and broke the grain, and in this manner the seeds were separated from the husks.

Separating the chaff from the grain was done on a windy day (hence the word "winnowing," which was once *windwian*). The threshed grain was scooped up in a winnowing tray and then tossed into the air in a windy part of the barn. The lighter chaff was carried off by the wind, while the heavier grain fell back into the tray. Another method of separating was by using a sieve that was called a grain "riddle."

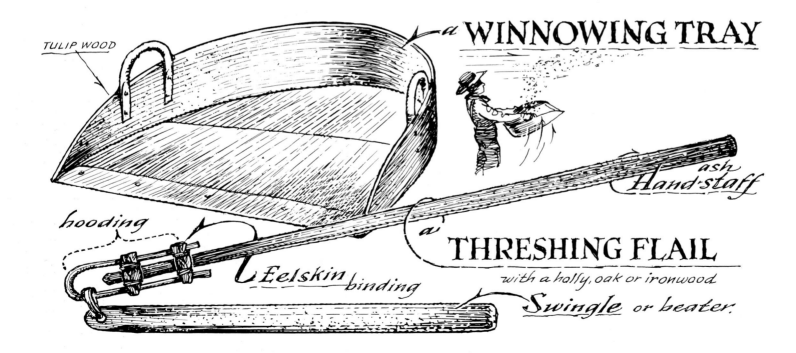

TULIP WOOD

a WINNOWING TRAY

ash Hand-staff

hooding

Eelskin binding

a THRESHING FLAIL
with a holly, oak or ironwood
Swingle or beater.

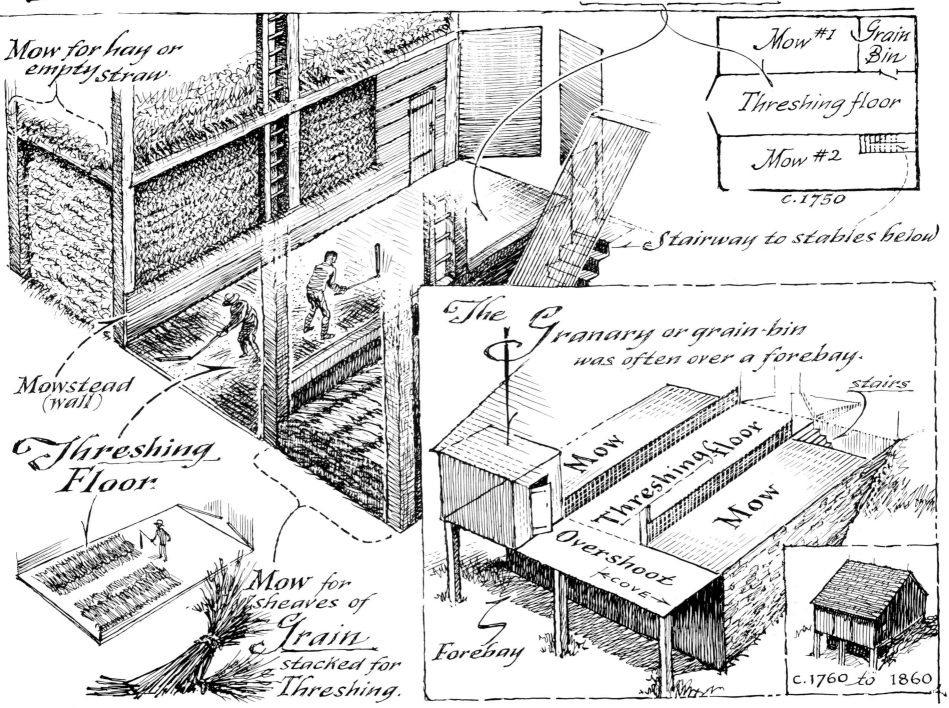

The OLD BARNS were built around THRESHING FLOORS.

Mow for hay or empty straw.

Stairway to stables below

Mow #1 | Grain Bin

Threshing floor

Mow #2

c.1750

Mowstead (wall)

Threshing Floor.

Mow for sheaves of Grain stacked for Threshing.

The Granary or grain-bin was often over a forebay.

Mow

Threshing floor

Mow

Overshoot ←cove→

stairs

Forebay

c.1760 to 1860

To Raise a Barn

On the opposite page, you can see some of the major steps in barn raising. The foundation was most often made without mortar (dry-wall construction), and the main girder—about twelve to eighteen inches thick—was the heart of the barn. The flooring in dwellings was usually left loose for a year or so before it was nailed down, but many barn floors were *never* permanently fastened.

The main framed and braced sections—usually four or five—were called "bents," and they were fashioned by the framer on the ground (as shown to the right) in preparation for raising day.

Neighbors for miles around came to help in a raising, which consisted of putting up only the framing. The siding and roofing work—which required much more time—was done by the owner and his helpers. At a barn raising, men came equipped with their own pikes and other tools. The framer—if one was hired to do the job—was seldom paid until after the raising, for if any corrections were necessary, he was expected to make them. This, of course, would delay the work of the raising crew, but any interruption from work was welcome, for there was always food, drink, and entertainment on hand.

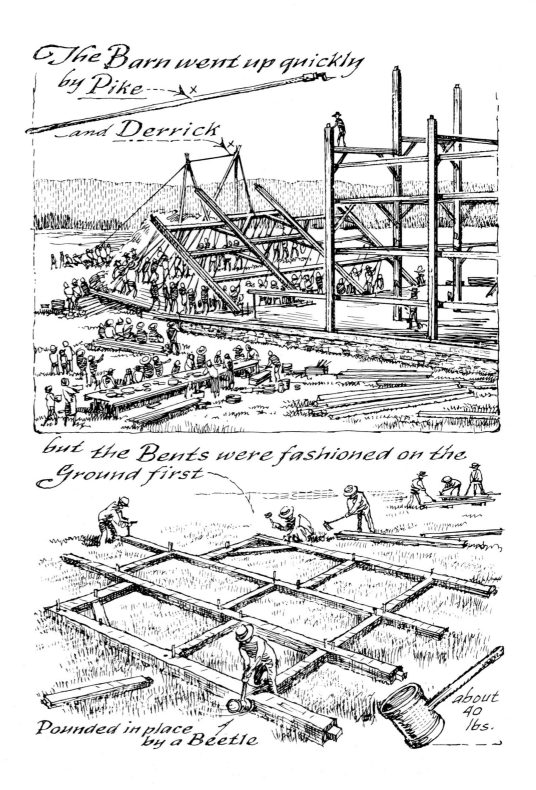

The Barn went up quickly by Pike and Derrick

but the Bents were fashioned on the Ground first

Pounded in place by a Beetle

about 40 lbs.

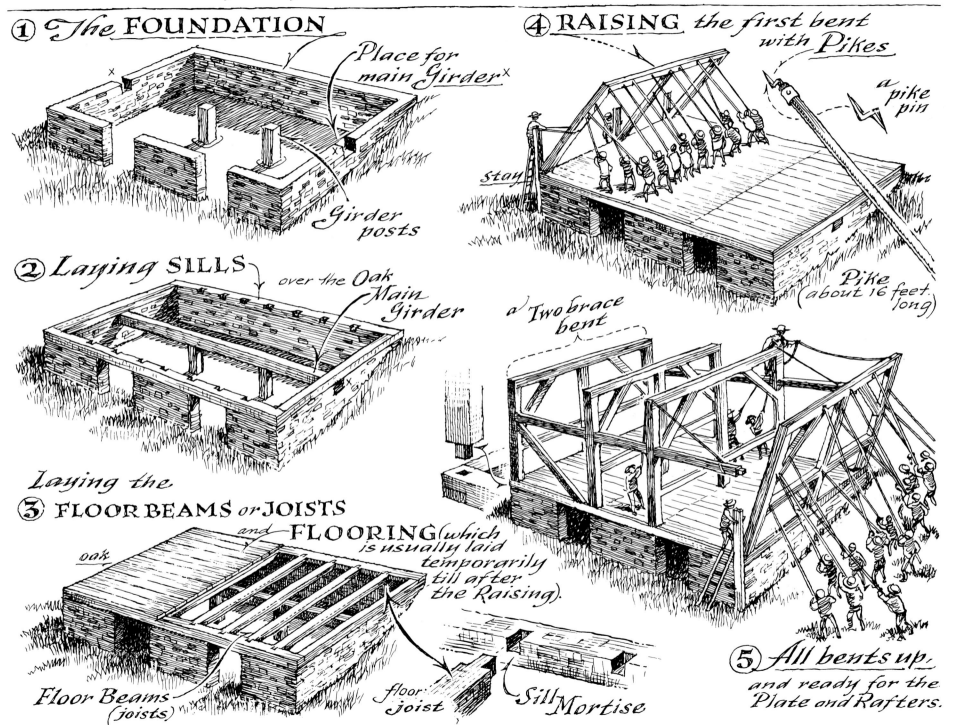

This is the way they built the barns ... ...

① The FOUNDATION

Place for main Girder ×

×

Girder posts

② Laying SILLS

over the Oak Main Girder

Laying the
③ FLOOR BEAMS or JOISTS

oak

and FLOORING (which is usually laid temporarily till after the Raising).

Floor Beams (joists)

floor joist

Sill Mortise

④ RAISING the first bent with Pikes

a pike pin

stay

Pike (about 16 feet long)

a Two brace bent

⑤ All bents up, and ready for the Plate and Rafters.

51

The UTOPIAN Shape

While we may never be sure where the idea for the round barn began, some evidence points to a religious origin. As American farming flourished in the 1800's, so did the association between the farmer and the church. Few men were nearer to God than those who worked close to the wonders of Nature. The farmer was also a model of sober living and hard work, seldom a changer of money, and incessantly a toiler to better the land.

There were many religious sects—most of whose members specialized in agriculture—who sought New World paradises. The Shakers, the Quakers, and the Holy Rollers all farmed with perfection as their aim. Each sect had its special farm architecture.

These sects were ever conscious of emblems, customs, and ways of life that set them apart from other church-going people, and the *circle* frequently became their theme—there were "sewing circles," "singing circles," and "praying circles." Farmers made circular designs on their barns, and their wives sewed circular designs on quilts. The Shakers used the circle in their "inspirational drawings" and invented the circular saw; they took delight in round hats, rugs, and boxes; and they made round drawer pulls and hand-rests as relief for their severely angled furniture.

There is a saying that the round barn was intended "to keep the devil from hiding in the corners."

Hancock, Massachusetts

1826

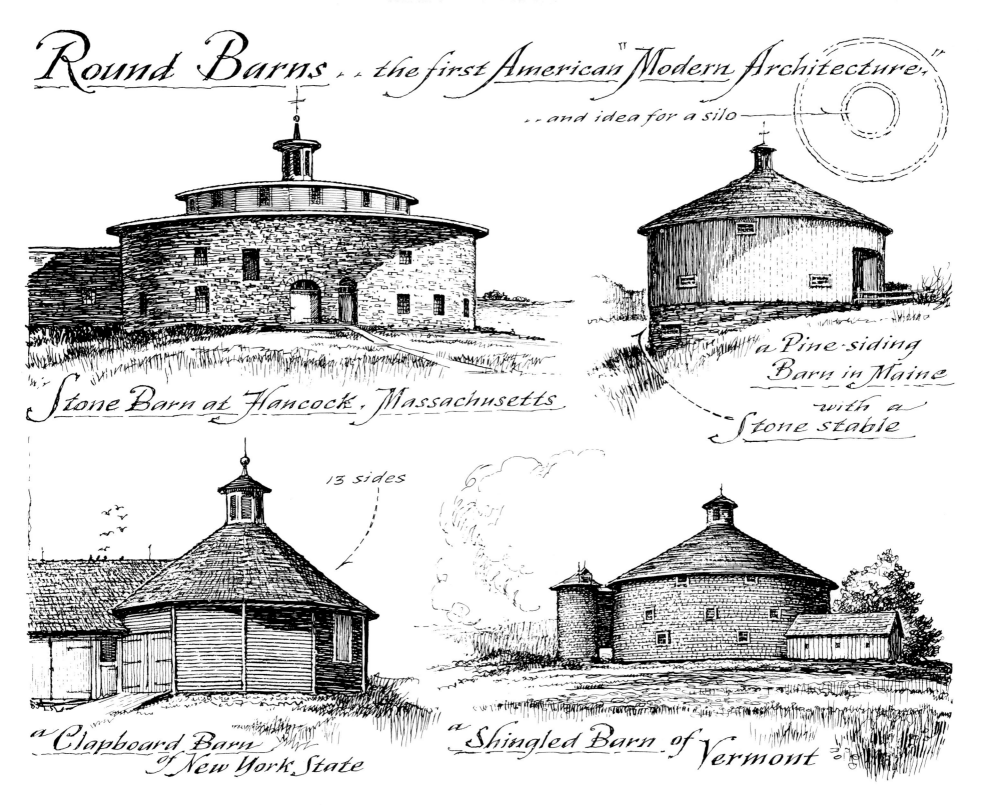

Round Barns .. the first American "Modern Architecture".

.. and idea for a silo

Stone Barn at Hancock, Massachusetts

a Pine-siding Barn in Maine with a Stone stable

13 sides

a Clapboard Barn of New York State

a Shingled Barn of Vermont

The "Factory of the Farm"

The first big, round barn of stone was built by the Shakers at Hancock, Massachusetts, in 1826. "The interior," they said, "was designed so that a great number of workers might be simultaneously engaged at their tasks and no person be in another's way." It had a fortlike security in its nearly yard-wide walls; it held fifty-two head of cattle; and there was an immense hay-storage area in its center. The center supports created a ventilating column that ended in a louvered cupola at the top.

On the circular driveway floor, which was fifteen feet wide, there was enough room for two hay wagons to pass each other and empty their loads into the center mow. Initially, this area was designed for threshing, too, but the "factory of the farm" soon proved too big for its own design and finally was used only as a cattle barn. Countless other smaller round barns were built, however, and many of them still remain—mostly in Vermont—and operate efficiently.

In an effort to create an American farm architecture based upon functional principles, the Hancock Barn (which partially burned in 1870) was a noble experiment. After nearly one hundred and fifty years, it can still be considered a symbol of modern design.

A small round barn of New Hampshire — *ventilator*

Hay is brought to the second floor and packed into the center silo. — *mow*

silo

stables

Hay is pulled below as needed

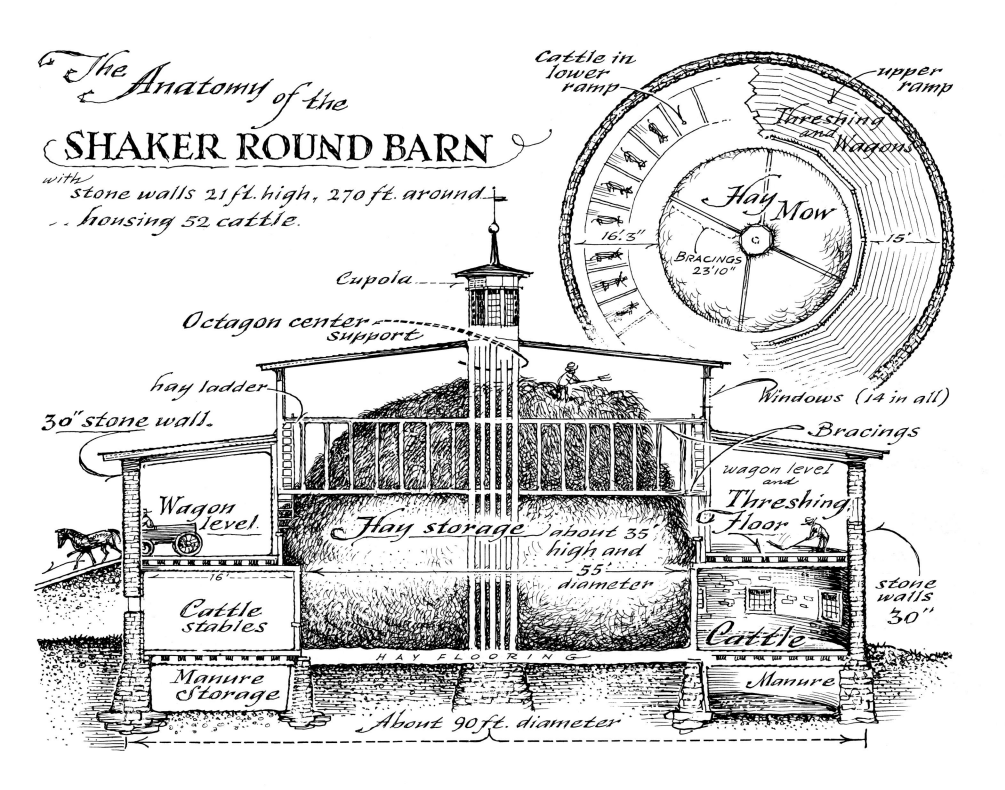

The Anatomy of the

SHAKER ROUND BARN

with stone walls 21 ft. high, 270 ft. around... ...housing 52 cattle.

Cattle in lower ramp

upper ramp

Threshing and Wagons

Hay Mow

16' 3"

BRACINGS 23' 10"

15'

Cupola

Octagon center support

Windows (14 in all)

hay ladder

30" stone wall.

Bracings

wagon level and Threshing Floor

Wagon level

Hay storage about 35' high and 55' diameter

16'

Cattle stables

stone walls 30"

Cattle

Manure Storage

HAY FLOORING

Manure

About 90 ft. diameter

The North Family Barn, New Lebanon N.Y.

With a cupola the size of a two-story cottage

Main Barn, original design.

c. 1885

c. 1860

Among the first to enjoy the American spirit of competition, the Shakers in New Lebanon, New York—only a few miles from the round barn at Hancock, Massachusetts—tried their hand at "a biggest of barns." About three hundred feet long and five stories high, this "North Family Barn" still stands. However, the flat roof proved leaky and was later changed as shown.

A similar barn (shown on next page), built at Lyons Falls, N.Y., no longer exists, but the legend of "The Big Barn" still lives.

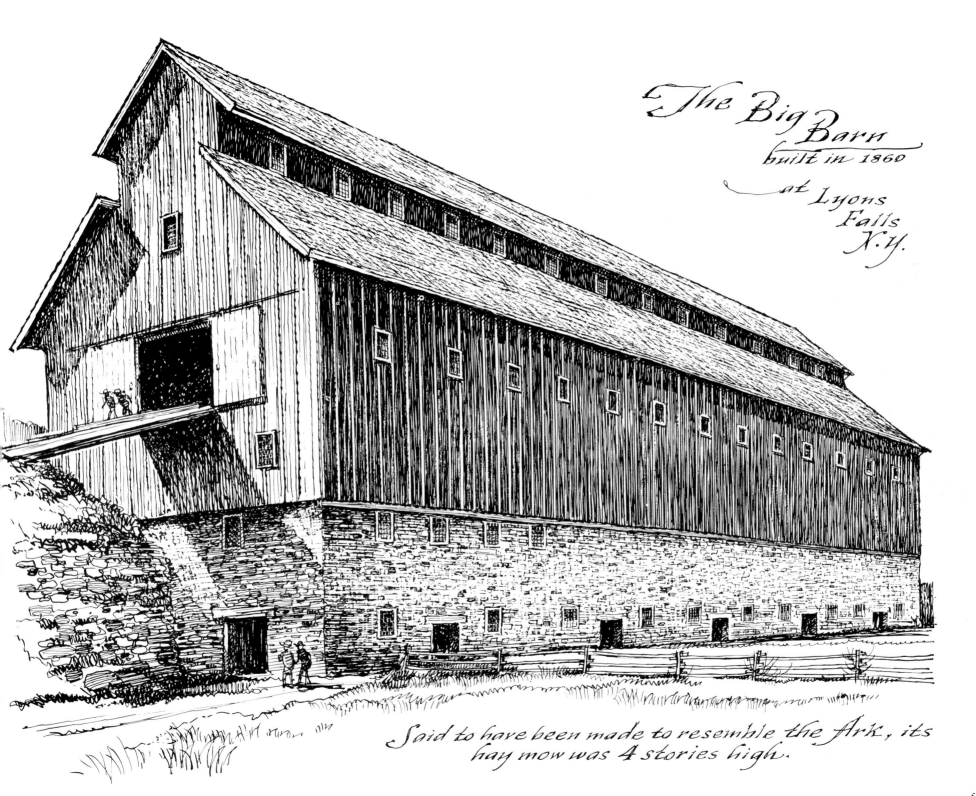

The Big Barn
built in 1860
at Lyons
Falls
N.Y.

Said to have been made to resemble the Ark, its
hay mow was 4 stories high.

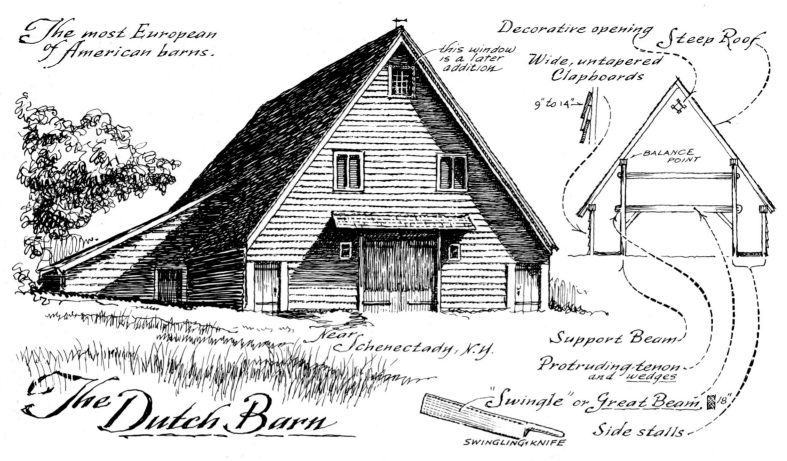

The most European of American barns.

Decorative opening · Steep Roof

this window is a later addition

Wide, untapered Clapboards

9" to 14"

BALANCE POINT

Near Schenectady, N.Y.

Support Beam

Protruding tenon and wedges

"Swingle" or Great Beam 18"

Side stalls

SWINGLING-KNIFE

The Dutch Barn

The true Dutch barn is the most "European" of American barns. It differs greatly from the Pennsylvania "Dutch" version, which, of course, is really German. Illustrated above are some of the features of the Dutch-American barn. Strangely enough, there are no such barns in The Netherlands; it was in America that the Dutch adopted and adapted these features from other European countries.

Only Dutch barns had finely adzed beams (in other barns the beams were broadaxed); the beams were not square, and they were usually ten inches wide and eighteen or more inches high. The tenons were thin and overlapped in a decorative curve, and they were *wedged* before pegging. The main or "great beam" was also called the swingle beam because it was narrow like a swingling-knife.

The long rafters—often measuring as much as sixty-five feet—were heavier at the bottom and were

made to balance perfectly on the "purlin-plate" before they were pinned at the peak. The "purlin-plate" is also European, and it brought the supporting uprights some ten to fifteen feet into the barn's interior, affording stall bays. The wall uprights carried no weight whatsoever.

The clapboards were really "lap-boards." Untapered, plain, wide boards, they were lapped one over the other. The wooden hinges were pinned directly into a door-frame mortise (A) or stapled in place (B).

Notice the similarity between these barns and the European barns shown earlier in the book. Actually, some of these Dutch barns were built in the 1600's, and the remaining ones in upper New York State may be the earliest examples of barn building in the United States.

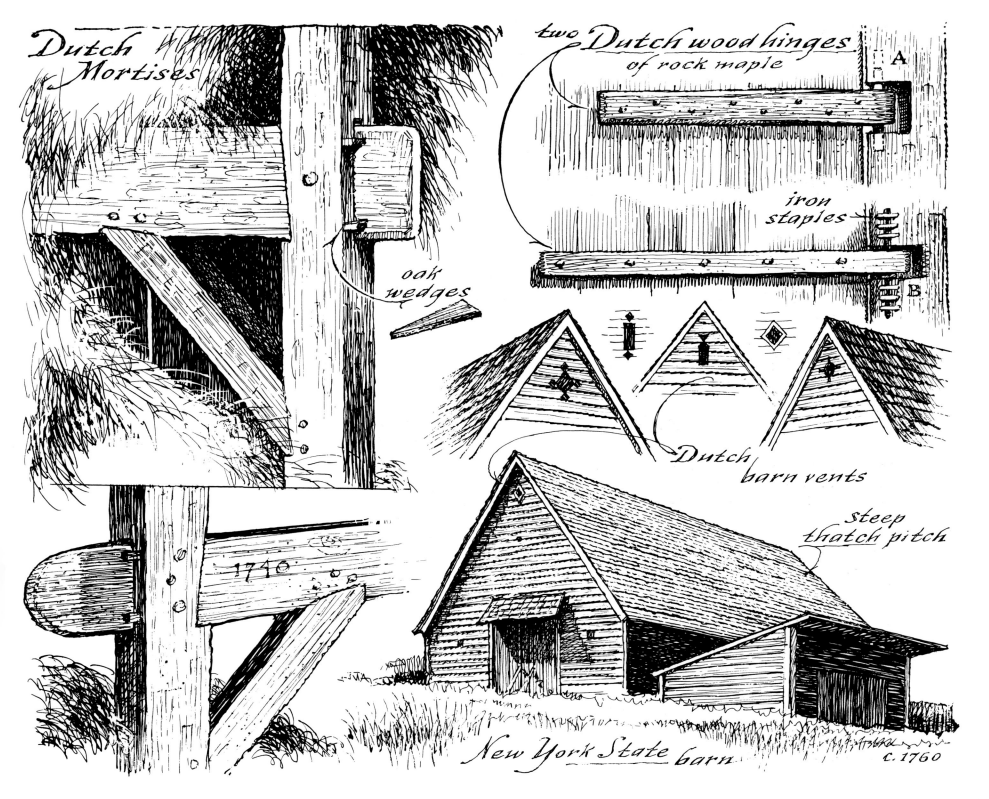

Dutch
Mortises

two Dutch wood hinges
of rock maple

A

oak
wedges

iron
staples

B

Dutch
barn vents

1740

steep
thatch pitch

New York State barn

c. 1760

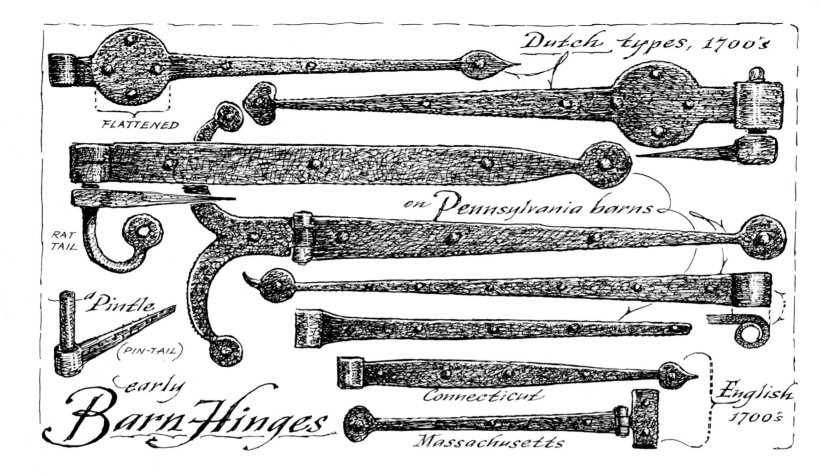

Dutch types, 1700's

FLATTENED

on Pennsylvania barns

RAT TAIL

"Pintle"

(PIN-TAIL)

early Barn-Hinges

Connecticut

Massachusetts

English 1700's

Generally speaking, hinges are almost never a clue to the age of a barn because doors took the hardest beating and over the years had to be replaced so often. For example, none of the original wooden barn-door hinges or leather granary-door hinges are known to remain.

Strap hinges were nailed to the barn doors and the wrought-iron nails were clinched or made "dead" (permanently fastened). In the same manner, all battened door nails were hammered down on their pointed sides. It is possible that the phrase "as dead as a door nail" derived from the custom of clinching nails, although this is only my idea and no one else has ever offered this explanation.

Many farmers made their own hinges in the farm forge—sometimes from wagon tires. Their designs were often copied from a favorite seen in another area—which accounts for a hinge of English design being found on a German barn, and so on. The Dutch were fairly consistent in their design, most often using the flattened disk. Since there were very few foundries in America, most of the earliest houses had hardware imported from overseas.

Pintles were sometimes notched to keep them from coming loose from the door frame; some had a "rattail" appendage that was nailed fast to the side of the timber.

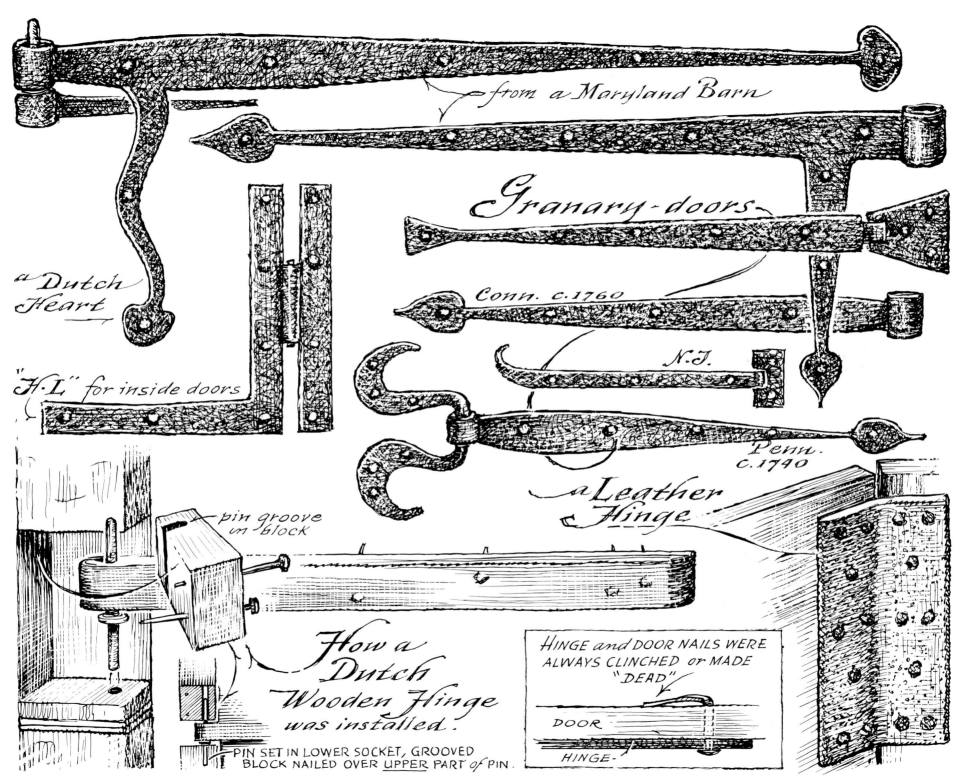

from a Maryland Barn

Granary doors

a Dutch Heart

"H·L" for inside doors

Conn. c.1760

N.J.

Penn. c.1740

a Leather Hinge

pin groove in block

How a Dutch Wooden Hinge was installed.

PIN SET IN LOWER SOCKET, GROOVED BLOCK NAILED OVER UPPER PART OF PIN.

HINGE and DOOR NAILS WERE ALWAYS CLINCHED or MADE "DEAD"

DOOR

HINGE

61

Silos

Many people associate the familiar upright silo with the early American farm, mainly because the remaining old farms generally have silos. But actually, the silos were added in later years. It comes as a surprise to many people to hear that *most* existing silos were built in the 1900's.

The first upright wooden silo is said to have been built in 1873 by Fred Hatch, in McHenry County, Illinois. It wasn't until two decades later that the idea was introduced in New England.

It may also come as a surprise to learn that the word "silo" once described a "hole in the ground," not an upright tower, and that until the late 1800's, silos were holes and ditches dug into the earth. Knight's *American Mechanical Dictionary* of 1876 states that a silo was "a large shallow ditch protected by thatch or boards from rain, sun, and air, and used as a store-pit for potatoes, etc." As shown below, the old-type farm silo was a copy of the Indian corn cellar, and it is very possible that the words "silo" and "cellar" have some connection.

The silo was the only farm building without a skeleton of framework; this method of building, initially a temporary measure, was continued for half a century. You can still see some old wooden silos, built only of a "skin" of light boards held together with wires and hoops, leaning this way or that, depending on the direction of the last storm it weathered.

an *Indian Corn Cellar* c. 1700
roof of boughs
Leaves
corn
Clay
Stones
10'

an early *Farm Cellar or Silo* c. 1800
Corn stalks
Beams
ashes or charcoal
Stone slabs

... leaning from forgotten winds, a discordant note
on many old farms,

... popping up like afterthoughts...
on the winter landscape.

Stone Silos

Wisconsin c.1879

c.1880

Connecticut c.1890

It has always been a mystery why New England is so lacking in stone buildings. A special wonderment arises when one sees occasional stone silos attached to old wooden barns. The answer is that the first stone silos were built during a time of a great lumber shortage. The late 1800's marked the time when charcoal-making for iron foundries had depleted many American woodlands, and stone was the only material available.

It is uncertain when the first stone silo was built in America: Michigan claims the credit with an 1875 date—the builder was named Manly Miles. Indeed, in Michigan and throughout Wisconsin are found some of the earliest and most outstanding examples of stone silos. Their thirty-inch thick walls were usually curved slightly to equalize internal pressures; and round stones were split in half, and their flat sides set outward.

In New England, the first upright silos were square, often built in one corner within the barn. Here the silo proved too hard to clean and, in addition, air seeped in and soured the silage, so the New England silo was moved outside and built in the rounded shape. The first stone models went as deep into the ground as they were high above the ground; they were squat, and like tiny European castle towers they lent an Old World mood to the American farmyard.

ROUND BARN

Obsolete and surviving only as a museum piece, this round stone barn was the wonder of American barn design a century ago. The solid stone water trough and the snow roller in the shed echo the past.

"Barn near Crandon, Wisconsin"

an Indiana Grain Barn

"Stovewood Masonry"

WOOD BLOCKS

Spring House in Montana

European settlers in the New World often copied the architecture of the Middle Ages, framing a house with massive timbers and then filling in the spaces between with brick, rubble or hay, and clay plaster. This is called "half-timbering." There are still many such dwellings and also a few such barns left in America. These barns are rarities and are not considered typical of the barn age, but when stacked logs were used in place of bricks, the effect is unusual enough to be worth recording here. In what was known as a "stovewood" wall, round and split cedar logs were laid crosswise in the manner of a well-stacked wood pile, and the spaces between were filled with lime mortar.

There are examples of "stovewood" walls in various regions where wood was plentiful and stones were scarce, but the design is best known as that of the German settlers of Wisconsin—where there are still a few such barns.

At the corners and ends of "stovewood" foundations, the wall was held in place by broadaxed, square oak logs; these were either notched and fastened log-cabin style or, after the manner of stone masonry, simply laid quoin style.

For Ventilation and for light too.

Dutch barn windows

Tip-window New Hampshire and Vermont

"German Hearts"

(men)

(women)

Except for the barn-door transom and "tip window" of New England, early barns seldom had glass windows. The tip window recalls the days when, if a farmer sold his house, he took his windows with him. When he built again, there might be one left over that he could use for the barn gable. Tipping the window brought it nearer to the peak of the gable.

Old stone barns frequently had vertical slits in their walls, and when bricks were introduced, the skillful Germans took full advantage of them in decorating their barns (see opposite page).

Most wooden barns with decorative cutouts have disappeared from the scene, and the only common cutout window that remains is in the outdoor privy. Parenthetically, it is interesting that the symbols for "sol" and "luna" were used to indicate the sexes in the days of early stagecoach inns.

The Dutch imitated church windows for the cutout decorations on their barns, and the Germans cut theirs in the shape of "flat hearts," stars, and tulips to ventilate and admit light. Both cutouts and brick windows are European devices, reminiscent of the days of castles when the window was a "wind-eye" or "wind-o"—an opening for both the wind and the eye.

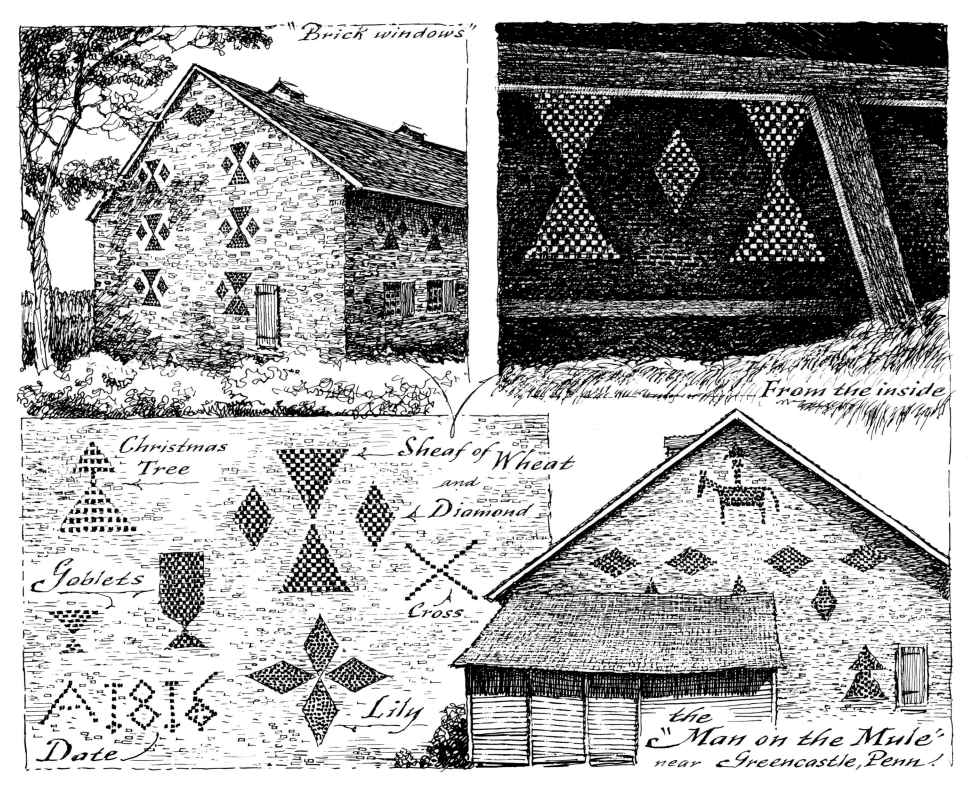

"Brick windows"

From the inside

Christmas Tree

Sheaf of Wheat and Diamond

Goblets

Cross

Lily

AD 1816

Date

the "Man on the Mule" near Greencastle, Penn!

Loop-holes and Louvers

Except for a few old Dutch houses that have loop-holes in the gable end of their attic walls, slit windows are rare in America. But most of the early stone barns had loop-hole ventilation. Where the name "loop-hole" came from isn't really known, but they were so called as far back as the fifteenth century, at which time they were designed as castle windows. The farmers called them loops or loop-holes when they first started putting them into their barns in the eighteenth century.

One hears that they were designed for spying and for shooting at Indians, but that is mostly romantic imagination. The flare or splay on the inside of a loop-hole window not only allowed a person to get a broader view, but also formed an aerodynamic situation that pulled the air out and prevented rain from entering. The inside splay was often white-washed to intensify the daylight that came in, and thus the inside of the barn was illuminated. Round holes were lined with brick and also splayed for better ventilation.

Louvers or "slat windows" were used on both stone and wooden barns, and most of the glassed windows you now see on old barns were originally louvered with wooden slats. They provided ventilation, but admitted very little light.

PENNSYLVANIA BANK BARN

Built into a bank with the animal entrance to the south, this stone barn with slit ventilators is typical of Pennsylvania German construction.

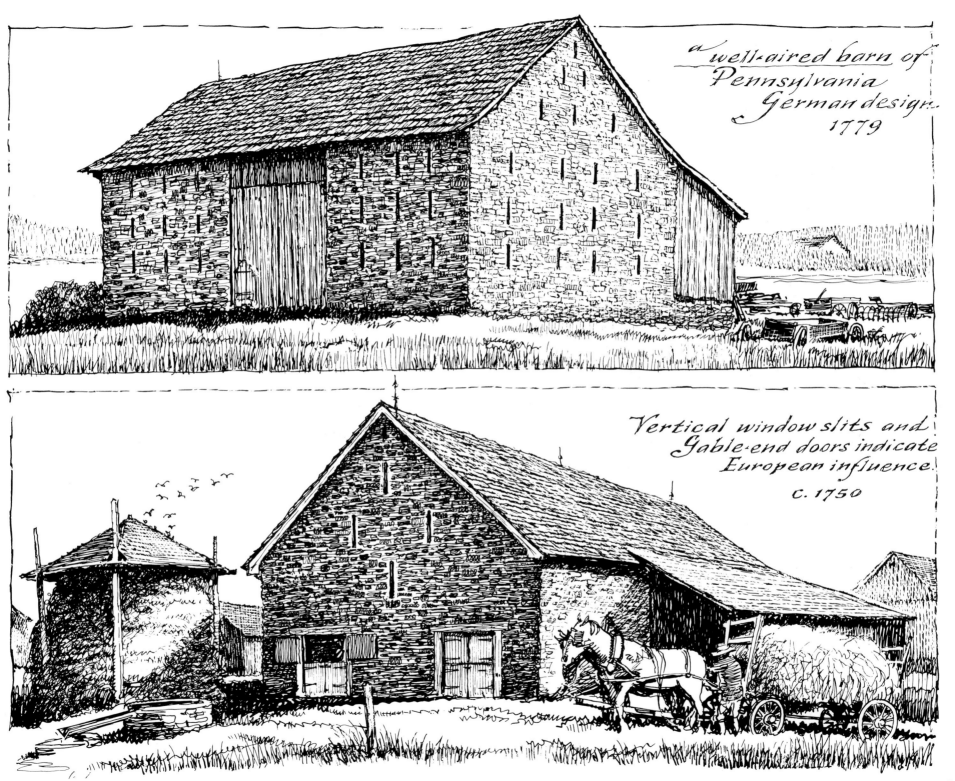

a well-aired barn of
Pennsylvania
German design.
1779

Vertical window slits and
Gable-end doors indicate
European influence.
c. 1750

two types of banked Connecticut cellars.

an Arched Root Cellar

sod

Hay

7 ft

10 ft

The Root Cellar

Made of stone with plaster on the inside, the root cellar of the 1700's was not dug down into the earth; instead, it was placed partly in a bank and then earth was piled around it. The floor was directly at ground level or a few inches above, and the door was wide enough for a wheelbarrow to be rolled through.

Just before frost, potatoes, turnips, cabbages, and carrots were taken to the root cellar and put in racks of dry hay. There, with the door toward the west, the greater bank to the north, and a building overhead to protect the cellar from the sun during warmer days, the effect was a comparatively even and mild temperature all year round. Being above ground, it was very dry.

A "cellar" in the old sense was not a hole in the ground, as we know it today; the word simply meant "storage place." When a cellar was completely beneath the surface, it was then known as a ground cellar.

The small root barn over a cellar held baskets, rooting tools, root cutters, and wheelbarrows, and there was room, too, for hanging late vegetables still clinging to their vines.

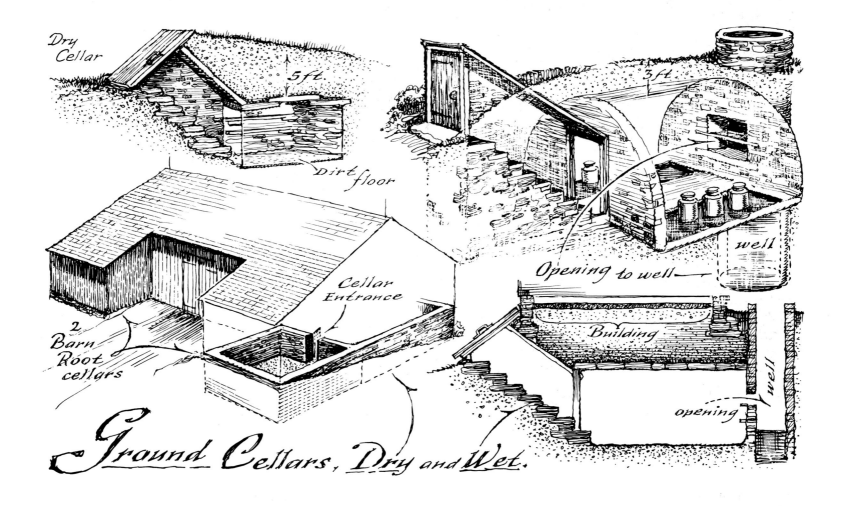

Dry Cellar

5 ft

Dirt floor

2 Barn Root cellars

Cellar Entrance

3 ft

well

Opening to well

Building

opening

well

Ground Cellars, Dry and Wet.

The cellar was a necessary part of every house, being the early cold-storage room for food; but the barn, too, often had its own cellars. Cellars were either *dry* or *wet*. The barn dry cellar was used mostly for root crops—turnips, potatoes, beets, etc.—but apples, cider, and dried fruits were also stored there. The wet cellar was built in connection with a spring or well; sometimes there was a constant flow of water through the cellar, and there were channels in which to place butter crocks and milk jars. It was still a cellar and not a springhouse, for the springhouse was built above ground. The drawing shows one wet cellar situated away from the barn and one directly under the building.

The barn dry cellar, which had an earthen floor, was for vegetables and fodder and had an entrance through the banked side of the barn foundation. Cellars like this were most often built of limestone; they had arched ceilings, and protruding stones in the wall to hold shelf boards.

Pennsylvania
c.1760

Overflow into cattle
trough

New Jersey Dry-wall
masonry

Ohio hillside
Spring House
1835

72

Spring Houses

1760

Inside a Spring House

Dug-out Spring

Door

Crocks standing in running water.

The location of a barn often depended upon the location of a spring, and the spring was sooner or later "housed"—both for cleanliness and for the storing of milk. The springhouse was a most attractive outbuilding, as it nestled in a ground hollow, protected from the summer sun by willows or other trees.

Unlike the wet ground cellar, the springhouse was built above ground and was generally equipped with one or two small, square windows. The overflow of many springhouses was carried directly to the barn by means of wood or lead pipe. The latter was made from hammered sheets of lead that were pounded into pipe form. Though crude, this was the first form of plumbing, and it is interesting that the Latin word for lead—*plumbum*—gives us our words "plumber" and "plumbing."

The sketches shown here have no particular architectural significance, but they do convey the picturesque qualities of the early farm springhouse. The springhouse roof was an ideal place for the growth of moss, and adding a sprinkling of earth accelerated the growth. Moss helped to keep the interior of the building cool. After a few decades, the shingles on the springhouse roof would be completely covered with plant growth.

Sugaring was first done with large iron pots

Vent Hood for steam

Oven Flue

Oven

Evaporator Pan

Sugar Houses

Sugarhouses were not built until the 1800's; prior to that time sap for maple sugar was boiled down in large iron kettles set in the open. The same kettles were used later in the season for boiling down apples for apple butter. For sugar, sap was collected from the maple trees in wooden troughs, then taken to a "sugar camp" where there were tents or rough shelters. Firewood had been stacked there the year before, in preparation for the spring event in the "sugar bush."

At the end of the 1700's, when tin evaporator pans and ovens were introduced, a better shelter was needed, and the "sugar cabbin" was born. The *cabbin* was where one slept, and since the sap boiled continuously for nearly a month—and needed constant watching—there were always bunks in the sugar *cabbin*. (This spelling changed to "cabin" about 1800.)

In the 1800's, the cabin became the sugarhouse—with a hooded opening in the roof peak and shelters for firewood. The roof was usually shingled with four-foot slabs of cedar or cedar bark. Although most sugarhouses of Vermont and New Hampshire were a distance from the farm, the New York and Massachusetts houses were often a part of the farm complex and were referred to as "sugar barns."

Vermont

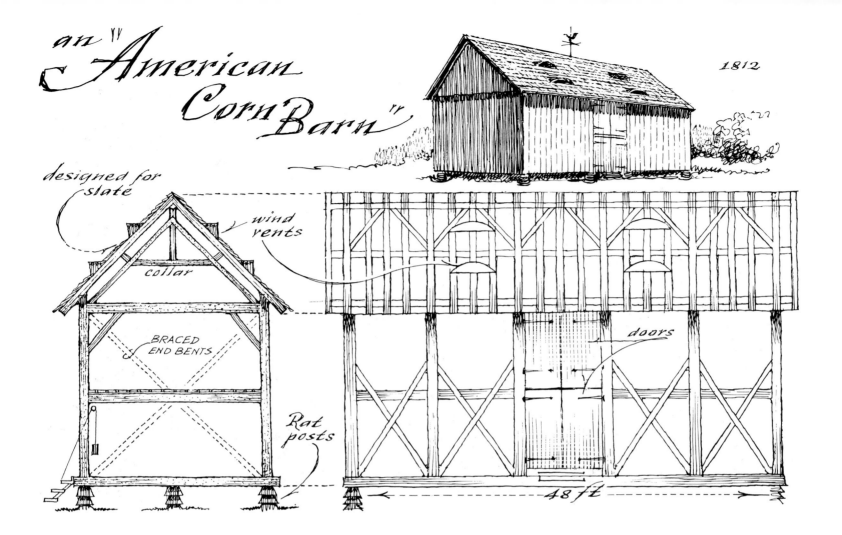

an "American Corn Barn"

1812

designed for slate

wind vents

collar

BRACED END BENTS

Rat posts

doors

48 ft

The sketch above is from the plan of a barn built in 1833 near Bernardston, Massachusetts. Surprisingly, the American corncrib is not American in origin at all—it was copied from an 1813 English plan. Accepted by some Americans (as evidenced by early paintings), this type was soon abandoned, but the small corncrib (shown on the opposite page) remained.

Impressed by our use of corn, Europeans presumed one would build a barn only for threshing and storing corn. Aware of "American windes," the plan above has vents "to keep the roof from blowing off." In England and elsewhere in Europe, corn had to be stored in barns that were erected on poles to protect the corn from house rodents. The first American farmers, fortunately, had little need for this protection.

Our small barns had four bents; this plan has six. Our small barns had no upper floor; this one has a slatted upper floor for storing corn in sacks. We drove wagons into our barns; this barn has a man-sized door and steps that were retracted by a weighted pulley arrangement.

the American Corn Crib

Drive-in Crib
N.J. 1830

Pan

Glass panes

c. 1700

New England
c. 1850

storm flap

retracting step

weight

Corn on the cob is seasoned and dried by allowing air to pass around it; this also keeps away mold. The corncrib is a slatted shed, protected from rodents by posts topped by pie tins, or sheathed with panes of glass. The first mention of a corncrib that I have been able to find is in an almanac for the year 1701. At that time, the crib was a square of alternately piled logs covered by a slanted roof. In the late 1700's, there were cribs built on posts, and many had one bin for soft corn and a larger bin for hard corn. The sidewalls always slanted outward at the eaves.

Large farms had several corncribs, because this was considered more efficient than one huge crib. On small pioneer farms, corn was dried in the dwelling attic or in a special "corn room"—usually over the kitchen.

The earliest farms also had cribs for the shelled corncobs. The cobs were prized for their oven ash, which was used in the smoking of meats, for quick kindling, and for dozens of other purposes. In many mountain farms, the drive-in crib was the beginning of barn architecture, for by adding doors in the driveway wall (see sketch above), you have the simplest American barn.

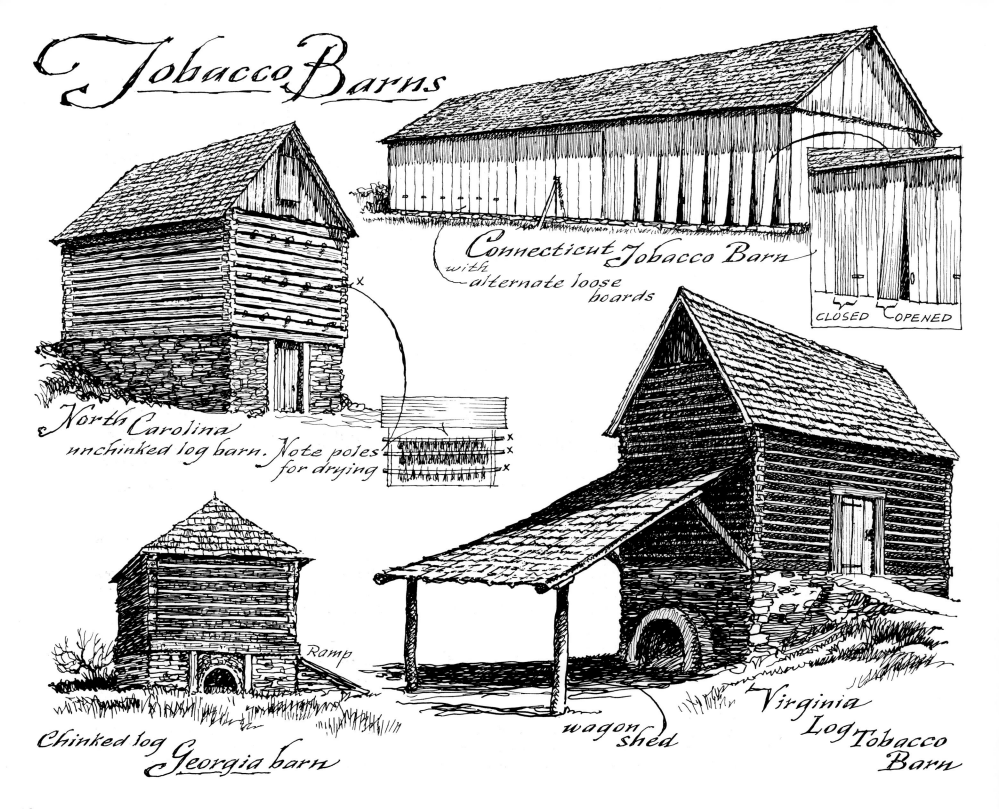

Tobacco Barns

Connecticut Tobacco Barn with alternate loose boards

CLOSED   OPENED

North Carolina unchinked log barn. Note poles for drying

Chinked log Georgia barn

Ramp

wagon shed

Virginia Log Tobacco Barn

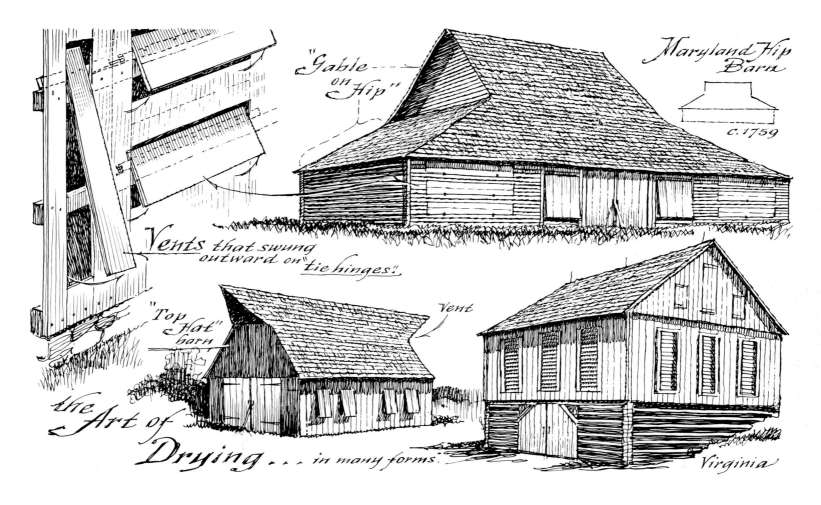

"Gable on Hip"

Maryland Hip Barn

c.1759

Vents that swung outward on "tie hinges"

Vent

"Top Hat" barn

the Art of Drying ... in many forms.

Virginia

More varied in design than other barns, the tobacco barn was one of the very first to be built in America. Tobacco *must* have a proper barn for drying. At first, tobacco was dried by air, then by the heat of a fire, and later by means of a metal flue. Ventilators and a roof that affords protection from rain are necessary for the air-curing of tobacco. According to location and custom, vents were hinged boards, slits of open log construction, or roof-peak louvers. Because fire-and-heat curing requires a tight enclosure, such barns were chinked and made as airtight as possible. Hardwood fuel, regulated with wet sawdust to burn slowly and produce smoke, was burned on the floor. The flue method uses a furnace and sheet metal piping running along the bottom of the barn.

Most southern tobacco barns were left unpainted. Some Kentucky barns, however, were painted, their vertical vent boards being daubed with a contrasting color. Southeastern Pennsylvania barns were often whitewashed; and until the middle 1800's, when red with white trim became popular, tobacco barns in New England were left unpainted.

Tobacco was often a small crop on the early farm. It was cured in the hay barn, either by hanging from loft poles, or by nailing leaves to the roof boards. This explains the hundreds of nails that extend down from many ancient barn ceilings.

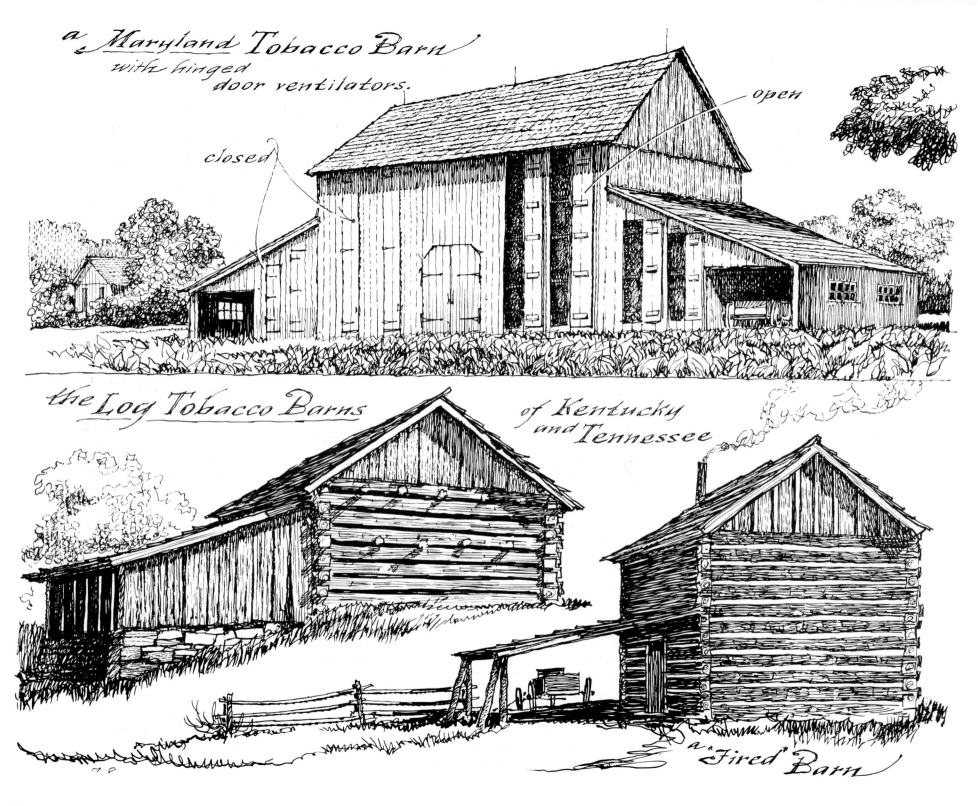

a Maryland Tobacco Barn
with hinged
door ventilators.

closed

open

the Log Tobacco Barns of Kentucky and Tennessee

a "Fired" Barn

80

a Striped Tobacco Barn

Kentucky c. 1875

a White Tobacco Barn of Pennsylvania

Roof Vent

Shown here are a few more examples of tobacco barns as they are found in different parts of America. The art of drying or "curing" tobacco was a delicate operation that challenged American inventiveness, and tobacco barns became a field of architectural competition, particularly in the area of air-conditioning.

The water content of tobacco, when it is harvested, reaches as high as ninety percent, and the method of drying always decides the quality and eventual worth of the crop. The heat or fire method is really more of a smoking method, and the barn for fire-cured tobacco is quite the same as a very large smokehouse; the result is a lingering smoky flavor in the tobacco, with a faint odor like that of creosote. Therefore, the slower method of drying, by controlled airflow, is the more popular method.

Although the many types of tobacco barns are suggestive of various rural scenes, their interiors are very much the same. They all have latticed, horizontal tier poles, spaced about four feet apart, which hold sticks of tobacco.

Tobacco and its barns are, if nothing else, completely American; no other crop, acre for acre, makes more money for the farmer—or needs more man hours to be produced. Although those built for curing tobacco were among the nation's first barns, they have less tradition and architectural lore than the livestock and grain-farming barns.

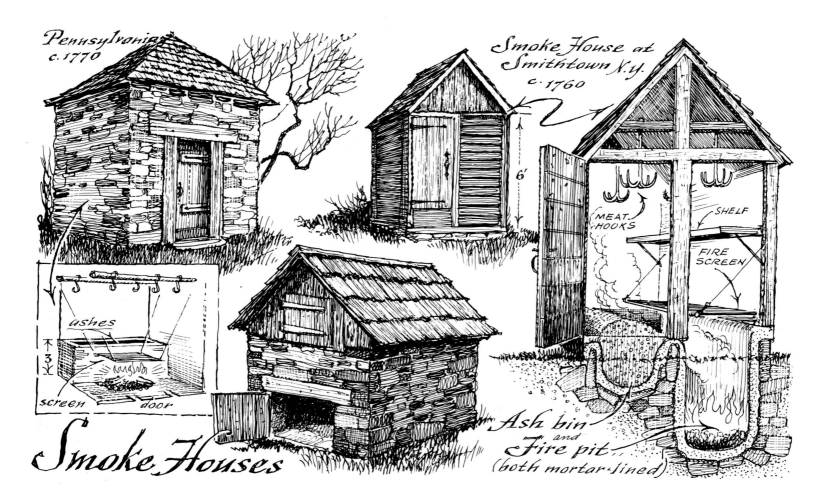

Pennsylvania c. 1770

Smoke House at Smithtown N.Y. c. 1760

6'

MEAT HOOKS

SHELF

FIRE SCREEN

ashes

3'

screen   door

Smoke Houses

Ash bin and Fire pit (both mortar-lined)

The pioneer American farmer had his small smoke rooms or smoke ovens connected to the fireplaces and chimney within his dwelling—from the cellar to the attic. Often they were on the second floor between two upper bedrooms. The attic smoke chamber did the best job, though, for there meat could be smoked with the least amount of heat. Meat was difficult to keep in those days, and smoking was more a preserving than a flavoring operation. In the 1700's, when the smoke ovens first went outdoors to become part of the barn complex, they often were large enough for a person to enter and hence were called smokehouses.

Many smokehouses were amazingly simple, being nothing but an airtight little house with a dirt floor. Hickory bark and corncobs were burned on the floor and meats were hung above—often protected by a tin screen—as far away as possible from the rising heat. Other smokehouses had ash bins where ashes would smolder till dead; meats were often buried directly in the ash bin for better preservation.

Although most smokehouses were completely airtight, some had vents at the sill line, and others had a chimney with an adjustable cover so that the fire could be better controlled.

This one was enlarged in 1860

a Drive-in Forge Barn

The Forge Barn

Pennsylvania c. 1790

The forge barn was more a tool house than a barn. There the constant repairing of broken gear meant that the coals were kept hot most days of the year. The only building in the farmyard with heat, the forge barn soon became the farmer's business office, storage place for tools, and little factory where new implements were made. Old files became froe blades, iron rods were wrought into nails and spikes, worn horseshoes were made into hinges, knife blades were cut into saws, and broken farm implements were reshaped into new ones. It was considered improper to waste anything when it could be salvaged or turned into something that would do another job.

You won't find forge barns on most of the remaining old farms since such work was given long ago to local blacksmiths. But a well-used bellows or some stray blacksmithing tools usually give evidence that this work was once done in the farmyard. In many cases, the forge barn was used as the dwelling while the farmhouse proper was being built. This explains why the few remaining forge barns—mostly built of stone—so often resemble tiny dwellings, with their windows and doors shaped differently from those of the other farmyard outbuildings.

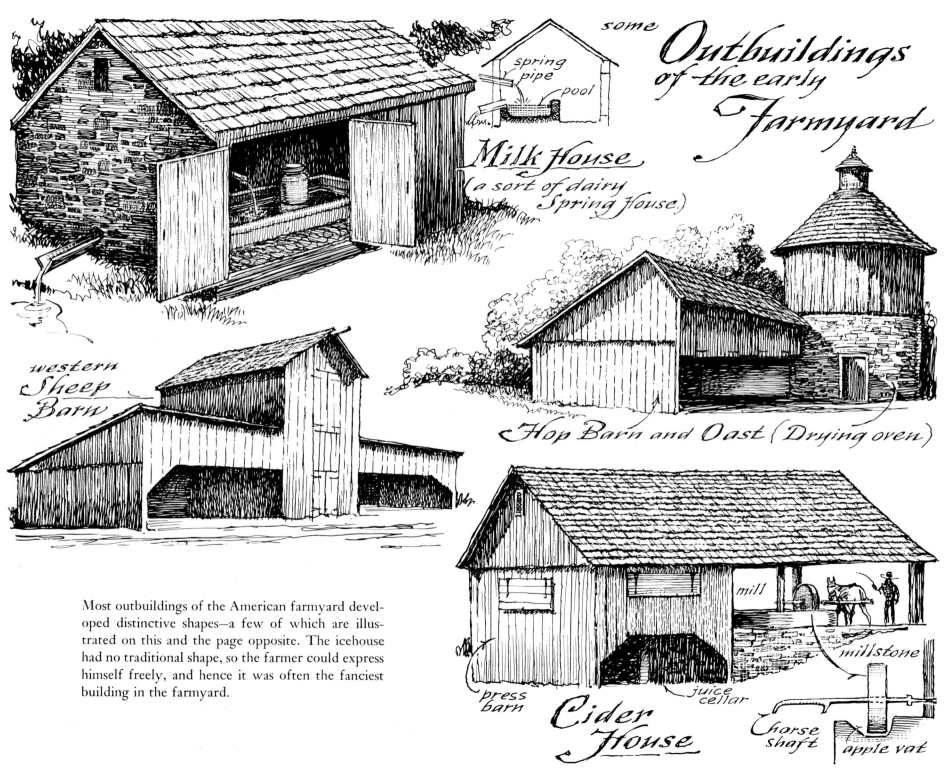

some **Outbuildings** of the early **Farmyard**

spring pipe

pool

Milk House (a sort of dairy Spring House.)

western **Sheep Barn**

Hop Barn and Oast (Drying oven)

mill

millstone

press barn

juice cellar

Cider House

horse shaft

apple vat

Most outbuildings of the American farmyard developed distinctive shapes—a few of which are illustrated on this and the page opposite. The icehouse had no traditional shape, so the farmer could express himself freely, and hence it was often the fanciest building in the farmyard.

Wash House
Virginia 1840

Outdoor Oven

c. 1820
Pa.

Portable Pig House
shaped so sow cannot roll on young.

Wagon Shed

Butchering Shed

Vents

Ice House

85

Barn Door Hoods

tulip swastika for bird's nest

Pennsylvania

Barn Doors

glassless "transom light" and a King size door-prop to last the winter.

Barn door Roller — 1840

With the earliest American barns, the large main door was a removable partition and often did not even have hinges. It was fastened shut all winter and taken off for the summer. Wagons were left under sheds, and the large threshing floor was kept completely free and clean.

The roller door, which most barns have been using for the past century, was introduced in the 1840's. (The patent was originally for "freight car doors.") An earlier sliding door ran on metal strips (as shown opposite) and needed constant greasing. But this was a large door with a smaller one within its framework (shown in sketch opposite), so the large door was infrequently used.

Earlier in this book you saw the long horizontal opening for barn illumination and ventilation, used during medieval days; this idea was repeated in America, but here it was placed over the door and had a hinged shutter. During the mid-nineteenth century, in New England, glass was added to these "transom lights" and the shutter was removed.

The leaning "prop-stick" that held the barn door shut is still used; nothing seems to work better. The sliding wooden bolt lock—because it had to be worked from the inside—was seldom used.

Nearly all small barn doors were once covered with a hood or "penthouse." Many of these hoods may still be seen in the Pennsylvania countryside. They were both practical and decorative and are worthy of revival.

GERMAN STONE BARN (SEED BIN)

This seed room was the "parlor" of a fine German stone barn. The hooded doorway and battened half-door are usually seen in Pennsylvania.

Door in a door

Sheathed Dutch door

Pennsylvania
c. 1750

Original shutter

Glass added
c. 1840

Connecticut
c. 1820

Pre-roller
Double-runner sliding door.

New Hampshire

mid
1700s

West Barnstable, Mass.

Connecticut Valley types

Ohio Ice House Cupola 1850

"Barn" "Cupolas"

"old Shaker Apple-drying Barn"

Derived from the word "cup," the first cupolas were domed turrets that were used for visual purposes, so barn cupolas are really misnamed. The typical early barn was without roof ventilators. The trend toward ventilation began in the Connecticut Valley when each farmer chose to express himself architecturally by designing an individual style of cupola.

Many godly farmers believed that lightning was God's will, and so they refused to use lightning rods. But scientists argued that the heat of fresh hay attracted electricity and a good ventilator would repel lightning. Therefore, farmers who refused "heathen lightning rods," accepted the cupola ventilators.

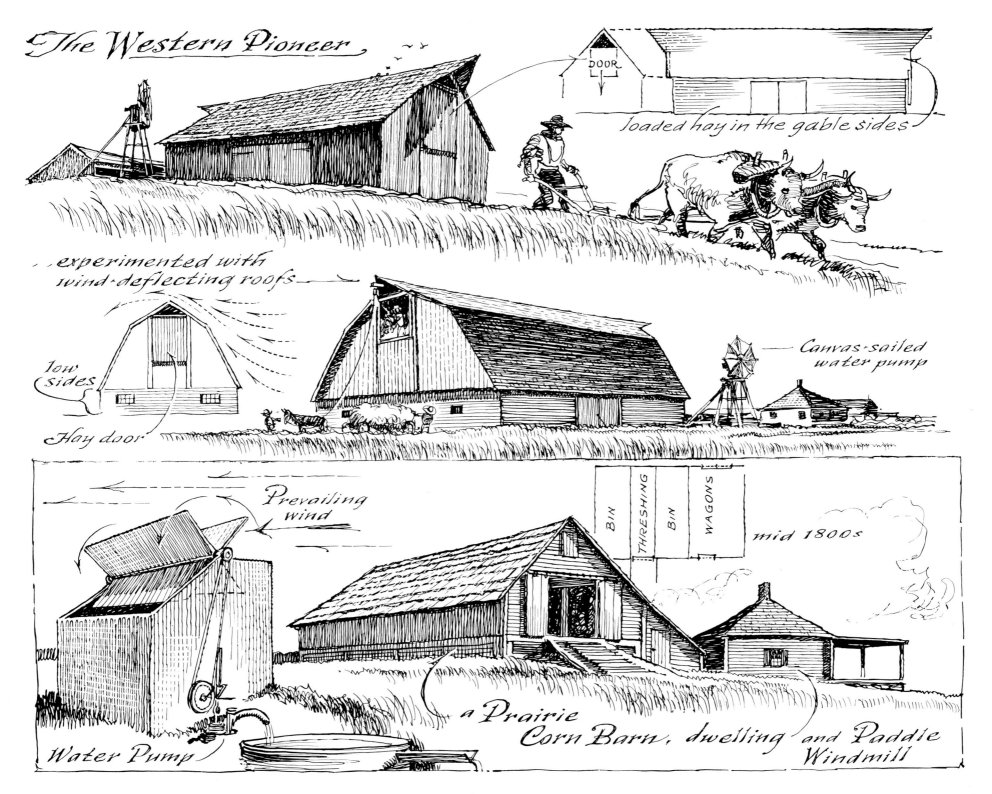

The Western Pioneer

DOOR

loaded hay in the gable sides

experimented with wind-deflecting roofs

low sides

Hay door

Canvas-sailed water pump

Prevailing wind

BIN THRESHING BIN WAGONS

mid 1800s

Water Pump

a Prairie Corn Barn, dwelling and Paddle Windmill

PRAIRIE BARN

With open-peak ventilation and lean-to livestock stall, this is the first type of large barn to appear in the West. Automatically directional windmill (with vane) came in the mid-1800's, the one with movable pole on a wagon wheel is from very early 1800's.

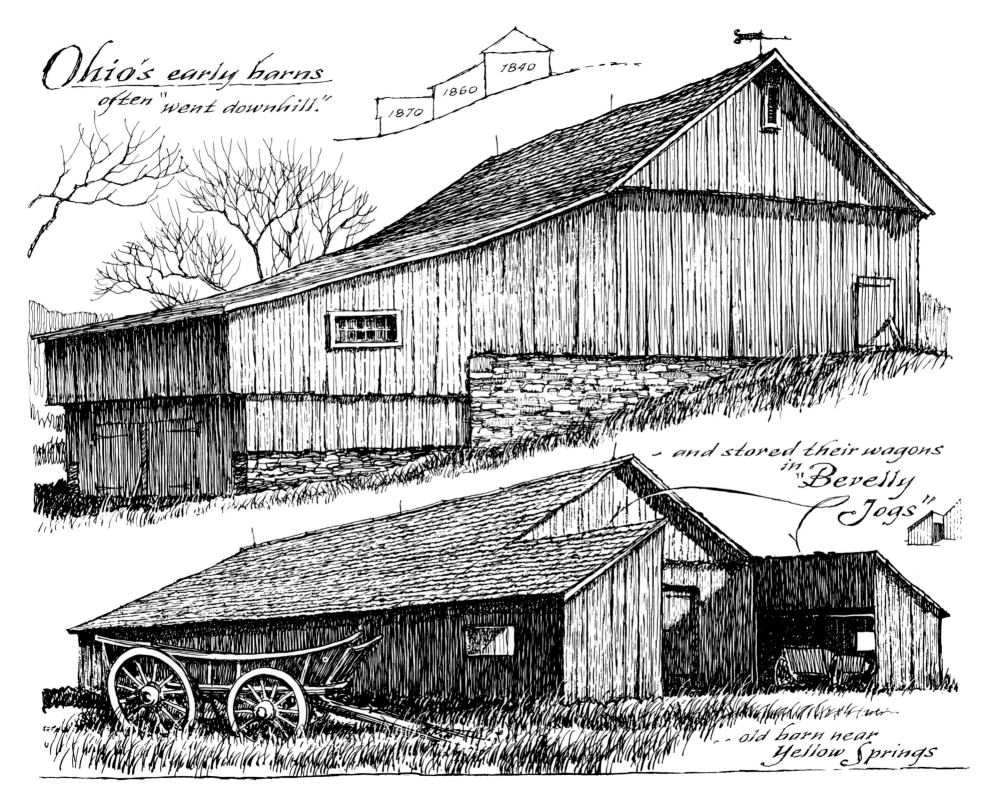

Ohio's early barns
often "went downhill."

1840

1860

1870

...and stored their wagons in "Beverly Jogs"

...old barn near Yellow Springs

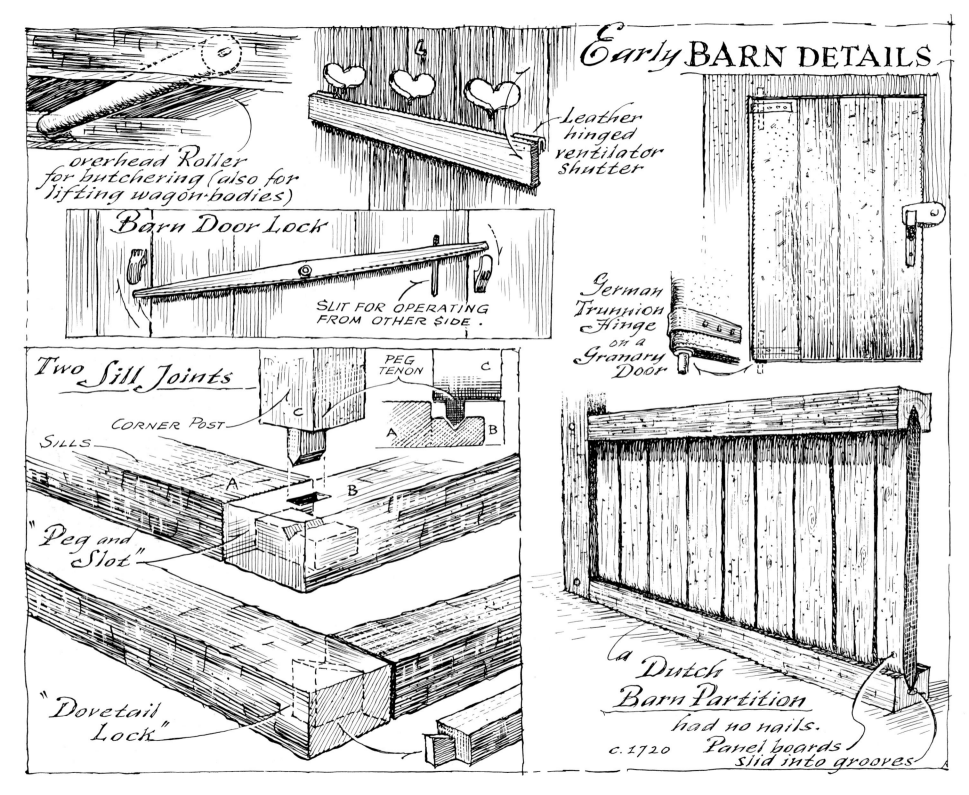

Early BARN DETAILS

overhead Roller
for butchering (also for
lifting wagon-bodies)

Leather
hinged
ventilator
shutter

Barn Door Lock

SLIT FOR OPERATING
FROM OTHER SIDE.

German
Trunnion
Hinge
on a
Granary
Door

Two Sill Joints

PEG
TENON

CORNER POST

C

A   B

C

SILLS

A   B

"Peg and
Slot"

"Dovetail
Lock"

a Dutch
Barn Partition
had no nails.
c. 1720   Panel boards
slid into grooves

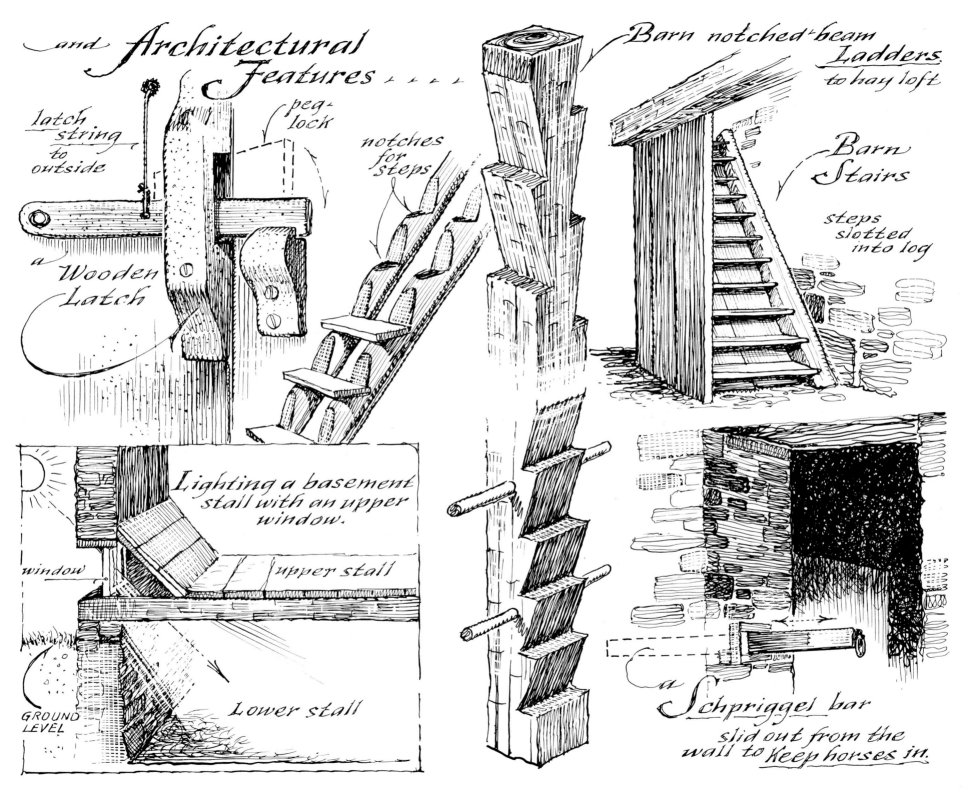

and *Architectural Features* . . . .

latch string to outside

peg-lock

notches for steps

a Wooden Latch

Barn notched-beam Ladders to hay loft

Barn Stairs

steps slotted into log

Lighting a basement stall with an upper window.

window

upper stall

GROUND LEVEL

Lower stall

a Schpriggel bar slid out from the wall to keep horses in.

BIBLIOGRAPHY

Piper, John, *Buildings and Prospects* (London: Architectural Press, Ltd., 1948)

Dornbusch, Charles H., *Pennsylvania German Barns* (Pennsylvania German Folklore Society, 1958)

Knight's *American Mechanical Dictionary* (1872)

Williams, H. L., and O. K., *Old American Houses* (New York: Coward McCann, 1957)

Shoemaker, A. L., *The Pennsylvania Barn* (Pennsylvania Folklore Society, 1959)

Raymond, Eleanor, A.I.A., *Early Domestic Architecture of Pennsylvania* (New York: William Helburn, Inc., 1931)

I have also consulted the writings of Henry Glassie, III, Richard W. E. Perrin, F.A.I.A., Dr. Fred Kniffen, Frank Wildung, Mary Earle Gould, and Wilbur Zelinsky.